CPI Theory

Continuous Planetary Interaction Theory

Complete in Three Parts

Tony Waterfall

2016

ISBN-13: 978-152363357
ISBN-10: 1523633352

Table of Contents

Foreword by Robert Currey

Scientific astrology has undergone a quiet renaissance in the last two decades. Yet in the second half of the twentieth century, Michel Gauquelin's evidence of a correlation between the visible planets at birth and eminence caused a great fanfare. His evidence also aroused a hard-line reaction from organised sceptical groups around the world. Their inability to refute his findings led to a concerted campaign of experiments and scientific papers that claimed to debunk astrology. Over time none of these sceptical results has stood up to critical analysis. Meanwhile, astrologers and scientific researchers have quietly continued to produce evidence supporting the simple thesis of a correlation between planetary movements and terrestrial life and events.

Nowadays, while the case against astrology is tenuous, one argument still carries weight: Astrology lacks a plausible mechanism. Of course, the lack does not justify dismissing astrology. But what concerns many scientists is that this absence does not concern most astrologers. It is likely that some symbolic astrological practices invoke forces that are unknowable or beyond our present knowledge or even that they rely on the placebo effect. Yet other astrological techniques with more conventional claims seem likely to be accounted for by a series of orthodox mechanisms. The challenge is to establish all the links in the chain from celestial to terrestrial. So when an astrologer like Tony Waterfall produces a model that proposes a plausible mechanism that accounts for the interaction between the planets, it should be of immense value to both astrology and to science.

I first met Tony Waterfall while I was researching the impact of orbital resonance between planets and the layout of the solar system. While my mechanism was based on gravity, Tony Waterfall was able to show how planets interact through electromagnetism. From the start I was impressed by his conjecture. In the intervening years, it has been exciting to learn of new scientific discoveries that support his CPI hypothesis. Auroral ovals may indeed be universal within the solar system. These are the visible evidence of invisible channels between planets. Tony Waterfall has been able to show that, far from being a collection of isolated and independent rocky spheres and gas giants, the solar system is a connected network.

Our first scientists, the ancient Babylonian astrologers, saw the wandering stars we know as planets as part of their pantheon. Their mythology narrated that these avatars for the gods interacted and talked to each other. While the anthropic mechanism of these ancient stargazers was evidently unfounded, three thousand years later Tony Waterfall's theory of Continuous Planetary Interaction demonstrates that their imaginary process may not be so far from the reality.

Robert Currey, BSc, DF Astrol.S, Cert.A*C*G Int., has been a professional astrologer since 1981. He founded the Astrology Shop in London in 1989. Currey consults, writes, lectures, conducts research, programs astrological software, for his company Equinox and specializes in Astro*Carto*Graphy (locational astrology). He is science editor for the Astrology News Service (ANS). In January 2016 Robert Currey was given a Lifetime Achievement Award at the 26th Indian Astrological Conference in Kolkata, India.

Foreword by Bruce Scofield

Astrology is a unique (and really quite strange) subject that in our present-day culture is highly polarizing. Those involved with it find that correlations between planetary positions and processes on Earth are blatantly obvious. Then there are skeptics who believe astrology is a pseudo science. It is near impossible to find somebody in this camp that actually knows something about the subject or has even tried it out. Their problems with it are (1) that it has not been explained in a way that is consistent with what is known in science and (2) it has not been rigorously tested. In regard to the second, there are tests but very few of them for one major reason – no money. In regards to the first, people involved with astrology are primarily interested in doing it, not getting to the science behind it. But here we have an exception to the rule.

Tony Waterfall has done something few students or practitioners of astrology have done – attempt to explain how it is that astrology could possibly work in terms of what is scientifically known about the physical universe today. That by itself makes his efforts worth something. What he has developed in his CPI hypothesis proposes that magnetic field patterns, specifically magnetic pole openings on planets, provide a means for charged particles from one planet to migrate to another. Although I personally would like to see more detail and evidence in certain areas, the basic idea of continuous planetary interaction as an astrological mechanism is certainly worth taking a look at.

Bruce Scofield, Ph.D., has been a general practitioner of astrology for 45 years, is an author of books and articles, and has long been a strong advocate for educational standards for astrology. He has taught evolution at the University of

Massachusetts and astrology for Kepler College. He has strong interest in Mesoamerican astrology.

Foreword by Robin Armstrong

CPI Theory (Continuous Planetary Interaction Theory) is a stimulating well referenced treatise on the scientific nature of astrological influence in modern astronomical terms. CPI Theory creates a new standard of perception and deduction in astrology, or at least a refreshing new approach. It will also influence traditional cosmologies that draw conclusions from astrology. It shows the astronomical essence of astrological or heavenly influences.

By using insights into the nature of auroral phenomena CPI Theory presents a logical argument for the nature of the interaction between planets and between other celestial bodies. Just as astrology presents lucid insights into the experiential nature of life and its relationships, so CPI Theory reveals lucid insights into the scientific nature of planetary influence. The scientifically minded will gain a clear validation of astrological planetary influence, and the astrologically minded will gain a brilliant affirmation of the scientific experiential foundations of astrology.

This is a work of multifaceted applications. I particularly like the idea that cycles of astrological significance could have a direct impact on the nature of micro cycles. This is, in my opinion, a great breakthrough filled with tremendous implications. That there could be seasonal cycles within the atom is a refreshing idea. That such implications could affect DNA presents a new gateway of insight. It also brings to the surface the idea of a subjective DNA. CPI Theory will encourage a redefinition of what science is.

Robin Armstrong, is a Canadian Astrologer, an international speaker and President of the RASA School of Astrology an internationally accredited school since 1994. He founded the Canadian Independent Astrologers Order in 1974 and was President of the IAO Research Library for thirty years. He was also a founding member, steering committee member, coordinator and advisor for the Association for Astrological Networking (AFAN).

Foreword by John Rutherford

In 1978, Dr. Bart Bok, perhaps the most prominent Astrophysical critic of Astrology at the time, told me personally, "I won't even take a look at Astrology until you show me the vehicle." Tony Waterfall has in his CPI Theory gathered what science has to show about that vehicle, at least the Astrophysical part.

In the world today Astrology has no unified scientific vehicle. Science has not found an Astrological vehicle because it hasn't looked. Scientists usually do not have the financial freedom or inclination to look into planetary influence and its Astrological implications. And in fairness they do not want to look since the answers may go against what they have been taught and the doctrine of the day. When something that favors Astrology is found the scientist sees no larger picture and the information goes into the files of saved data.

Around us what works, what affects many species, and what provides information for guidance in many ways is the electromagnetic flux in our extended environment. Only recently was mammalian cytochrome 1 discovered, the pigment in the eye that helps give animals, including humans, a magnetic orientation.

There are some details in CPI Theory that I see differently than Tony Waterfall but there are enough areas that we agree upon that I encourage you to read CPI Theory and come to your own conclusions about continuous planetary influence.

John Rutherford, *is an "Astrologer at Large" reading horoscopes since 1971. He co-authored a program on Astrology for the H. R. MacMillan Planetarium in 1979 and* *lectured on Contemporary Astrology at the Southam Observatory in 1981. He lectured on Astrology for the School of Architecture, UBC, 1982 and was Astrology Host for "In Touch" daily TV program, CFRN, 1985. He writes a weekly Astrology Newsletter.*

Editor's Preface

CPI Theory is a reference work in three parts that has a table of contents listing 36 chapters and 58 subchapters, each with a descriptive title. There are 671 sources listed at the end of each chapter. Each of the three parts contains an Abstract, Introduction, Summary and Conclusion. There is a Bibliography that is recommended reading. There are also three separate indexes at the back for ease of navigation. The Astrological Index, Solar System Index and the Electromagnetic Phenomena Index have been created to give the reader an informative door into the vast amount of information contained within CPI Theory.

CPI Theory has been peer reviewed by several leading astrologers from Canada, U.K., USA, and India. CPI Theory was written over a four year period in three separate parts. CPI Theory Part One was completed December 2013 and published as The Theory of Continuous Planetary Interaction in the 2014 Research Journal of the National Council for Geocosmic Research. CPI Theory Part Two was completed December 2014 and published in a serialized format in the 2015, 2016, and 2017 Research Journals of the National Council for Geocosmic Research. CPI Theory Part Three was completed March 2016. All three parts are contained in this edition and have been re-edited and updated to reflect recent advances at time of publication.

Thanks to Susan Harkins for help setting up the index formulas and troubleshooting. Thanks to Erika Waterfall for help with a multitude of editing issues.

Author's Preface

I started writing CPI Theory to unfold a lifetime of astrological belief that the planets of our solar system affect our lives here on Earth. I know this is easy to say but where is the proof? Science has the proof and that is what is laid out within these pages as Continuous Planetary Interaction Theory (CPI Theory).

CPI Theory Part One was inspired by Professor Syun-Ichi Akasofu through his highly informative book Exploring the Secrets of the Aurora. CPI Theory Part Two was inspired by Astrologer Bruce Scofield, PhD., who cautioned against confusing both the reader and myself. This caused me to read many source books to find an understanding and a path. CPI Theory Part Three had a life of its own with a strong foundation based on the peer support of many Astrologers including; Robert Currey, Bruce Scofield, PhD., Robin Armstrong, John Rutherford and Jagdish C. Maheshri, PhD.

Please enjoy the ride to a destination that is exhilarating.

PART ONE

Abstract

All of the astrological planets (Sun, Moon, Mercury, Venus, Mars, Jupiter, Saturn, Uranus, Neptune, Pluto and Earth) should have a surrounding interactive sphere. These spheres have magnetic poles. At the magnetic poles there can be auroral ovals. One auroral oval sits above the attracting magnetically negative pole and the other above the discharging magnetically positive pole. The discharge travels through open space. It follows the path of least resistance and greatest attraction to cover vast distances while creating continuous waves in the yielding dark and cold medium of interplanetary space.

The discharge energy from a planet is drawn in by Earth's magnetic field and enters Earth's environmental sphere. Some energy is visible light. Neutrons and protons become atoms the universal building blocks that interact with Earth's atmospheric environment. Electrons from interplanetary space hit Earth along a ring shaped oval. This creates an electrical discharge powered by the solar wind that interacts with the magnetosphere. Solar wind and the effects of solar discharges upon the solar wind create a pressurizing mechanism for the continuous auroral flow that will contain travelling matter and unique signatures from all planets in our solar system.

On Earth, the planetary discharges arrive channelled by the magnetic field and are drawn through the auroral ovals. Through electromagnetic commonality and through biophysical processes these planetary discharges are acted upon by human beings.

Introduction

Though mainstream science recognises the impact of the Sun and the Moon on Earth's biosphere, it is not widely accepted that the planets have any influence on terrestrial events and ultimately human affairs. This is arguably the main reason why astrology remains on the fringe of science.

The classical astrological elements of earth, water, air and fire are viewed from a scientific perspective by Philip Ball, a Professor of Physics who says, "The classical elements are familiar representatives of the different physical states that matter can adopt. Earth represents not just soil or rock, but all solids. Water is the archetype of all liquids; air, of all gases and vapours. Fire is a unique and striking phenomenon. Fire is actually a dancing plasma of molecules and molecular fragments, excited into a glowing state by heat. It is not a substance as such, but a variable combination of substances in a particular and unusual state caused by a chemical reaction." [a1]

Professor Syun-Ichi Akasofu (now retired) Director, International Arctic Research Center says, "I am convinced that new thinking is needed to solve long-standing unsolved problems...The new generation of scientists are encouraged to challenge the present paradigms and advance our understanding of electromagnetic phenomena around the Earth, in interplanetary space, and the heliosphere...In this age of infinite specialization, the importance of synthesis should be emphasized more now than ever...My dream of a grand synthesis is to bring solar physics, interplanetary physics, magnetospheric physics, and upper atmosphere physics together in terms of space weather research." [1]

The answer to planetary involvement may already be available. NASA and others have collected and continue to collect large amounts of data. [2] Perhaps the data can get organized into a clear statement of fact that leaves no doubt that planetary influence is alive and well. [3]

The model that I am proposing can be summed up in twelve statements.

1. Each planet is surrounded by an interactive sphere. [4]
2. The sphere is surrounded by a magnetosphere, some faint, some strong. [5]
3. Each sphere has two auroral ovals one above each magnetic pole. [6]
4. The magnetic polarity of the poles under the ovals will change with time as the poles flip magnetic polarity. [7]
5. One auroral oval sits above a magnetically positive pole and the other above the magnetically negative pole. [8]
6. There is a flow through the ovals with its strongest direction dictated by the magnetic polarity of the pole below. [9] [9a]
7. The flow consists of electrons, protons, neutrons and more particles yet to be identified. [10]
8. The flow at the oval is constant like the flow of water from a tap that is never turned off. [11]
9. The size and flow through the auroral oval differs from planet to planet. On Earth where quality study has taken place the auroral ovals are found to be of considerable size. The auroral oval above Earth's negative magnetic North Pole is 6,400 kilometers in diameter while the auroral oval above Earth's positive South Pole is 6,000 kilometers in diameter. [12]

10. The average rate of magnetic flow through Earth's incoming auroral curtain that forms the outer edge of the auroral oval is 1,000,000 megawatts. [13]
11. If a cross-section of the entrance flow at Earth's incoming auroral oval was monitored it should show a composition made up of contributions from each of the astrological planets. [14]
12. At any given moment an astrological signature from each of the planets should be entering Earth's atmosphere and the environment of Earth's biosphere. [15]

I have tried to lay out the workings for CPI Theory Part One in four chapters (1) Planetary Sphere (2) Auroral Oval (3) Space Transference (4) Planetary Dispersion.

I do not claim that CPI Theory Part One is the answer to how astrology works, but it is my hope that this theory will show at least one mechanism for continuous planetary interaction with life on Earth.

Sources for Introduction

[a1] 2004, Ball, Philip, *The Elements: A Very Short Introduction* Oxford University Press. Kindle Edition. (L 359)
[1] 2007, Dr.Syun-Ichi Akasofu (now retired) Director, International Arctic Research Center, *Exploring the Secrets of the Aurora*, Second Edition, page v, Volume 346, Astrophysics and Space Science Library.
[1a] 2009, Akasofu, Syun-Ichi, *The Northern Lights* page 264, Alaska Northwest Books
[2] 2013, Earth's auroral oval is monitored by the NOOA POES satellite and by the DSMP satellite and many others. A real time display can be found at http://sd-ww.jhuapl.edu/Aurora/ovation_live/ovationdisplay.cgi?pole=N&type=E

And at the Geophysical Institute, University of Alaska
http://www.gi.alaska.edu/AuroraForecast
[3] 2013, September 22, Daily Galaxy, *Evidence of Extraterrestrial Life Found in Earth's Atmosphere, Challenged.*
http://www.dailygalaxy.com/my_weblog/2013/09/claims-of-extraterrestrial-life-found-in-earths-stratosphere-challenged.html?utm_source=feedburner&utm_medium=feed&utm_campaign=Feed%3A+TheDailyGalax
[4] 1999, August 16, NASA, *Planet in a Test Tube*, paragraph 9
http://science.nasa.gov/science-news/science-at-nasa/1999/msad16aug99_1/
[5] 2015, NASA-JPL,
http://genesismission.jpl.nasa.gov/science/module4_solarmax/solarmax_planets.html
[6] 2007, SunEarthPlan, *Extra-terrestrial Aurora*,
http://www.sunearthplan.net/6/310/Extra-terrestrial-aurora
[7] 2013, August 5, NASA, *The Sun's Magnetic Field is About to Flip*
http://science.nasa.gov/science-news/science-at-nasa/2013/05aug_fieldflip/
[8] 2006, Georgia State University, *Hyperphysics, Aurora*,
http://hyperphysics.phy-astr.gsu.edu/hbase/atmos/aurora.html
[9] 2005, *Earth's Auroras Don't Mirror*,
http://www.nasa.gov/vision/earth/lookingatearth/dueling_auroras.html
(9a) Stanford, *Visualizing the Ionosphere from 150Kkm*
http://solar-center.stanford.edu/SID/educators/ionosphere_metadata.htm
[10] 2012, September 20, Journal of Geophysical Research,
http://onlinelibrary.wiley.com/doi/10.1029/JA093iA07p07441/abstract
[11] 2013, 1994, Aurora Activity Extrapolated, NOAA POES
http://www.swpc.noaa.gov/pmap/ Cambridge Journals
http://journals.cambridge.org/action/displayAbstract;jsessionid=73751E4B52C847DFFA46F1170347A9DB.journals?fromPage=online&aid=5415848
[12] 2012, *Diurnal Variation of the Auroral Oval Size.*
http://onlinelibrary.wiley.com/doi/10.1029/JA084iA09p05319/abstract
[13] 2009, Akasofu, Syun-Ichi, *The Northern Lights* page 139, Alaska Northwest Books

[14] 2013 Real time, Space Weather Updates by SolarHam,
http://www.solarham.net/data.htm
[15] 2013 Real time, Space Weather Updates by SolarHam,
http://www.solarham.net/data.htm

Planetary Sphere

The concept of spheres around planets has been postulated since Ptolemy and Copernicus and goes back to at least 1250–1300 when the Latin *sphaera* for globe and the Greek *sphaira* for ball were in use. [16] Spheres are quite common on Earth where we have a magnetosphere, atmosphere, exosphere, ionosphere, thermosphere, mesosphere, stratosphere, troposphere, heterosphere, homosphere, biosphere, anthroposphere, noosphere, hydrosphere, cryosphere, pedosphere, geosphere, lithosphere, asthenosphere, mesosphere, mantlesphere, and a coresphere.

Dr. Anne Hofmeister, Research Professor at the Department of Earth and Planetary Sciences says, when talking about her recent research paper, "Existing models for planetary accretion assume that the planets form from the dusty 2-D disk, but they don't conserve angular momentum. It seemed obvious to me to start with a 3-D cloud of gas, and conserve angular momentum. The key equations in the paper deal with converting gravitational potential to rotational energy, coupled with conservation of angular momentum…In all cases, the process is gravitational accretion of these cold, 3-D clouds making things contract and spin out, and that's where the energy comes from. It's all happening in very cold temperatures, in 3-D instead of 2-D." [17]

The nebula hypothesis [18] of star formation accepts that a star is formed by the gravitational collapse of a pocket of matter within a giant molecular cloud.

It seems quite possible to place a sphere-like cloud around a planet and add an auroral oval above each magnetic pole. This infers two holes within the sphere that through an electromagnetic process facilitate movement of mass from outside the sphere to inside the sphere and then from inside the sphere back to outside the sphere. Mass arises through the interaction between two particles. Interaction happens at the auroral oval.

The sciences accept [19] that most of the planets in our solar system have auroral ovals but based on today's current data it is said that Mars, Venus and Earth's Moon have no auroral ovals. It is possible that the sciences will discover active auroral ovals on Mars, Venus and the Moon in the very near future.

Sources for Planetary Sphere

[16] 1985, Van Helden, Albert, *Measuring the Universe*, pp. 37, 40
[17] 2012, Hofmeister, Anne, Research Professor, Department of Earth and Planetary Sciences, when talking about her recent research paper February 27. Washington University in St. Louis
http://news.wustl.edu/news/Pages/23466.aspx
[18] 2006,
http://csep10.phys.utk.edu/astr161/lect/solarsys/nebular.html
Cambridge University Press. Brandtner, Wolfgang and Klahar, Hubert *Planet Formation: Theory, Observations, and Experiments*
[19] 2011, Space Telescope Science Institute Nebula Hypothesis, Aurora Alive, *Find the Aurora on Other Planets*
http://auroraalive.com/multimedia/autoformat/get_swf.php?videoSite
=aurora&videoFile=aa_find_the_aurora.swf++&videoTitle=Find+the+
Aurora

Auroral Oval

Many people are aware of the aurora borealis also known as the northern lights. [20] This dynamic visual light display is the outer visual edge of the inflowing auroral oval cascading 100 kilometers into the atmosphere of Earth as colored chemical rain traveling through a magnetic field that is motivated by an electrical current. The auroral oval acts as a generator producing up to 10,000,000 megawatts of electrical power while a 1,000,000 ampere current flows through the auroral curtain. [21]

Professor Syun-Ichi Akasofu says, "The Aurora can then be understood as the only visible manifestation of the electrical discharge processes that are powered by the dynamo. Thus auroral activity and geomagnetic disturbances are only different manifestations of an enhanced dynamo power." [22]

Light is emitted when charged particles trapped in the magnetic field interact with atoms of atmospheric gases. The colour depends on which atoms in the Earth's atmosphere are emitting the light. Plasmas and electrons traveling the field lines spiral in and hit the atmosphere, generating light whose colors depend on which chemicals – oxygen (red and green light), nitrogen (blue light), and others - are excited in this process. When auroral light displays occur, about two thirds of the time, they cause the Earth's atmosphere to expand slightly. [23]

There are two auroral ovals above planet Earth and CPI Theory hypothesises that there should be two auroral ovals above each of the astrological planets: Sun, Mercury, Earth, Jupiter, Saturn, Uranus, Neptune and Pluto. Auroral ovals may be found on the Mars, Venus and Earth's Moon no matter how faint or distorted.

On Earth at this moment in our history the north auroral oval sits above the magnetically negative north pole that induces material through the auroral oval towards the magnetic pole and into the atmosphere of Earth.

Earth's southern auroral oval sits above a positively charged pole disbursing material out to re-circulate [24] and to travel beyond Earth linking up with the space currents. I propose that this re-circulation and interplanetary space interaction happens on all planets.

Each planet through its auroral ovals will disburse and receive material. The flow rates and contents will differ but each planet will have a unique signature.

Sun: Coronal holes are regions of low-density plasma on the Sun that have magnetic fields that open freely into interplanetary space. During times of low solar activity, coronal holes cover the north and south polar caps of the Sun. During more active periods coronal holes can exist at all solar latitudes but they may only persist for several solar rotations before evolving into a different magnetic configuration. Ionized atoms and electrons flow along the open magnetic fields in coronal holes to form the high-speed component of the solar wind. [24a]

Mercury: Dr's. C. T. Russell and J. G. Luhmann say, "Mercury is of great importance to those studying planetary magnetic dynamos and to those studying planetary magnetospheres. Its importance to the magnetic dynamo problem stems from its being the smallest and most slowly rotating planet with a presently active magnetic dynamo. Its importance to the physics of planetary magnetospheres stems from its lack of a dynamically important atmosphere or ionosphere. Currents generated by the solar wind interaction, which usually close in the ionosphere, cannot close in the same way at Mercury as they do in other planetary magnetospheres. It is thought therefore that the Mercury magnetosphere may be more strongly coupled to the solar wind than is the case for other planetary magnetospheres." [25]

Venus: Venus' atmosphere is made mainly of carbon dioxide with thick clouds made of sulphuric acid and other atmospheric gases. It has almost no magnetic field but a magnetosphere has been detected. Venus is thought to have an induced magnetosphere where the solar wind interacts directly with the planetary ionosphere. [26]

Moon: The air on Earth's Moon's is a mix of argon-40, which seeps out of the ground due to radioactive decay in the lunar interior while elements such as helium, sodium, and potassium are blown off the lunar surface by solar wind. I propose that this discharge will leave the lunar sphere via an auroral oval that sits above the positive pole.

Based on data from India's Chandrayaan-1 Luna Probe, a mini-magnetosphere had been observed on Earth's Moon and other magnetospheres might be found on the Moon perhaps around impact craters. [27]

Martin Wieser, a Senior Scientist at the Swedish Institute of Space Physics says, "The discovery of the Moon's first mini magnetosphere opens the door for finding magnetospheres on smaller bodies, even asteroids." [28]

Mars: Mars produces a weak magnetic field that partly deflects the solar wind leaving the sciences puzzled as they think that Mars once had a thick atmosphere, but they also think it may have been blown away by the solar wind. NASA has sent an orbiter to Mars called MAVEN for Mars Atmosphere and Volatile Evolution that reached Mars in 2014. I propose that if Mars once had a thick atmosphere then it is likely that it was magnetized and had a magnetosphere and thus auroral ovals. Bruce Jakosky, Principal Investigator for MAVEN says, "MAVEN will determine if loss to space was the most important player in driving Martian climate change." [29]

I propose that if Mars once had a magnetosphere, it is possible that it has not disappeared but it is faint or has been distorted to the point that it is currently unrecognizable.

A University of California at Berkeley, press release says, "Though Mars lacks a global protective magnetic shield like that of the Earth, strong localized magnetic fields imbedded in the crust appear to be a significant barrier to erosion of the Mars atmosphere by the solar wind." [30]

Astronomer, Astrologer and Astrophysicist Dr. Percy Seymour says about data from Mars Global Surveyor, "The new data show that where the localized surface magnetic fields are strong, the ionosphere reaches to a higher altitude, indicating that the solar wind is being kept at bay." [31]

If Dr. Anne Hofmeister and the nebula hypothesis is right, then Mars, Venus and Earth's Moon can have a cloud with a polarized magnetosphere and auroral ovals. I propose that these planets might have an electromagnetic disbursement and capture mechanism and that is through their auroral ovals.

Jupiter: Jupiter is a gas giant that rotates very rapidly generating strong magnetic fields in its interior that are surrounded by a giant magnetosphere that contains moons like Ganymede that possess its own magnetic field creating a mini-magnetosphere within Jupiter's much larger magnetosphere. [32]

On Jupiter the solar wind plays a lesser role than on Earth. Electrified sulphur atoms spewed from active volcanoes on Jupiter's moon Io, follow Jupiter's magnetic field through its auroral oval until they crash into Jupiter's upper atmosphere, setting it aglow with color.

Rapid spinning inside Jupiter's ionosphere and magnetic field causes a rotational motion that dominates the magnetosphere. Millions of amperes of electricity flow between the magnetosphere and ionosphere creating currents powerful enough to generate an auroral oval around each pole.

Saturn: Saturn's auroral oval is similar to that of Earth with auroral crowns that are created by fluctuations in the solar wind. [33] Saturn's upper atmosphere is dominated by hydrogen gas causing it to glow deep ruby red when excited by incoming electrons, unlike the more common green auroras produced by oxygen molecules here on Earth. The inflowing auroral oval at Saturn's negative pole takes in large quantities of sulfur emitted from its moon Titan.

Saturn like Jupiter is a gas giant that rotates very rapidly generating strong magnetic fields in its interior that are surrounded by a giant magnetosphere that contain moons. Saturn's moon Enceladus has an auroral oval.

Uranus: Uranus is mostly made of rock and ice. The polar regions are the warmest places on the planet. Due to the unique rotation of Uranus, scientists are currently not sure which of the poles is negatively charged. [34]

Neptune: The composition of Neptune is assumed to be close to Uranus' and made mostly of hydrogen and helium gases. The winds on Neptune can reach up to 2,000 kilometers per hour, the fastest winds out of all the planets in the solar system. [35] These winds should put strong pressure on the outgoing auroral oval.

Pluto: Although little is known about the atmosphere of Pluto, scientists guess that it is made of nitrogen and a mix of carbon monoxide and methane and likely has a magnetic field. [36] Existing data should be re-examined incorporating new data provided by the 2015 flyby by the New Horizons mission.

Earth: Earth's northern auroral oval has a continuous curtain around its outer edge. This curtain is half a kilometer thick and extends for many thousands of kilometers somewhere in the order of 20,000 kilometers. [37] Within this curtain, you will find the northern lights with their varying intensities caused by a high-vacuum electrical discharge within the mixture flowing through the auroral oval via the auroral curtain; an electric dynamo releasing into Earth's atmosphere.

The entrance surface of Earth's auroral oval contains about 3 million square kilometers while the auroral curtain contains about 10,000 square kilometers of entrance surface. The auroral curtain extends downward 100 kilometers giving it a volume of one million cubic kilometers of dynamic interaction between the magnetic field of Earth, the interplanetary magnetic field and the solar wind.

The auroral curtain hangs down into Earth's atmosphere reaching an elevation 60 miles off the surface of the Earth. Another 10 miles closer to the surface of the Earth, noctilucent clouds circle the globe. [38]

The area below the Earth's northern auroral oval is a region of heating and strong winds that carry the interplanetary inflow into the ionosphere and upper atmosphere. The auroral oval is fixed with respect to the Sun while the Earth rotates under the auroral oval. The geographic pattern of the auroral oval discharge changes as Earth rotates explaining along with solar wind, why auroral light can at times be seen over vast areas. [39]

The Earth's auroral oval is found at the intersection of the Van Allen radiation belts and the ionosphere and is the boundary between the highly stretched almost vertical magnetic field lines of the polar cap and the more normal field lines at lower latitudes. [40] Plasmas with electrons traveling the field lines spiral in and hit the atmosphere, generating light whose colors depend on which chemicals - oxygen, nitrogen, and others - are excited in this process.

On Earth, the sound of the auroral oval is intense emitting radio waves over wide frequencies in the range of 500 – 1,600 kilohertz. When auroral activity increases, more galactic radio noise is absorbed into the ionosphere. [41]

The causes of the aurora have been well studied by Dr.Syun-Ichi Akasofu, when he was Director of the International Arctic Research Center. He and others have brought the implications of the auroral oval and Earth's magnetic field into view as Earth's geomagnetic field becomes one with the interplanetary field. The auroral oval size will vary with the strength and direction of the interplanetary magnetic field.

I propose that the auroral oval is a common mechanism that every astrological planet should possess. With eleven astrological planets (Earth included), there should be at least twenty-two auroral ovals in our solar system breathing in interplanetary chemicals and building blocks and discharging planetary influenced complexities back into our solar system. Time will only find more complexities of interaction as more auroral ovals are located.

The intake and discharge interaction happens with varying degrees of pressure applied to the auroral ovals by the continuous never stopping solar wind made up of electrically charged particles. The almost constantly changing solar wind affects the voltage of the auroral generator and changes the electric and magnetic fields around Earth. There are several measuring devises being used today, but no one seems to be looking at a cross section of the flow and its composition. [42]

At speeds of 36,000 kilometers an hour, a stream of particles comprised mostly of electrons and protons, flows out from the Sun. As it flows out, the solar wind, the result of the hot corona expanding into space accelerates and can reach speeds as high as 3 million kilometers an hour. Ion cyclotron waves made of protons circle around the Sun's magnetic field. [43] They accelerate the solar wind and cause it to heat up as it blows into cold space. Chemical elements of the solar wind such as hydrogen, helium and heavier ions, blow at different speeds and change temperature with direction.

The solar wind usually reaches Earth with a velocity around 1.5 million kilometers an hour. [44] The intensity of the magnetic field hitting Earth is measured at 2–5 nanoteslas. Earth's surface field is usually 30,000–50,000 nanoteslas. During magnetic storms, flows can be several times more intense while the interplanetary magnetic field may also be much stronger. [45]

Earth's magnetosphere is formed by the impact of the solar wind on the Earth's magnetic field that forms an obstacle to the solar wind, diverting it, at an average distance of about 70,000 kilometers forming a bow shock [46] 12,000 kilometers to 15,000 kilometers further upstream. The width of the magnetosphere abreast of Earth is typically 190,000 kilometers, and on the night side a long magnetotail [47] of stretched field lines extends to great distances.

The flow pattern of magnetospheric plasma moves from the magnetotail toward the Earth, around the Earth and back into the solar wind through the magnetopause on the day-side taking with it outflow from the southern auroral oval.

The magnetosphere is full of trapped plasma as the solar wind passes the Earth. The flow of plasma into the magnetosphere increases with increases in solar wind density and speed. Some magnetospheric plasma travel down along the Earth's magnetic field lines and lose energy to the atmosphere in the auroral zones. Magnetospheric electrons accelerated downward by field-aligned electric fields cause the bright aurora features. The un-accelerated electrons and ions cause the dim glow of the diffuse aurora.

Sources for Auroral Oval

[20] 2001, *Northern Lights, The Science, Myth and Wonder of Aurora Borealis.*

Photography by Calvin Hall and Daryl Pederson with Essay by George Bryson.

[21] 2009, Akasofu, Syun-Ichi *The Northern Lights* Alaska Northwest Books, pages 77, 145 and 158.

[22] 2007, Akasofu, Syun-Ichi, *Exploring the Secrets of the Aurora* (Second Edition, page xxv.)

[23] 2000, 2008, University of Alaska, Fairbanks, *Why Are There Colors in the Aurora*, http://ffden-2.phys.uaf.edu/211.fall2000.web.projects/Christina%20Shaw/Aurora Colors.html and http://www.space.com/6229-earth-atmosphere-breathes-rapidly-thought.html

[24] 2012, *Recirculation of Plasma Sheet Particles Into the High-latitude Boundary Layer*, Journal of Geophysical Research http://onlinelibrary.wiley.com/doi/10.1029/98JA02392/abstract;jsessionid=87CF65B232AD6D5FCA014B01CDAFC6CA.f02t02?deniedAccessCustomisedMessage=&userIsAuthenticated=false

(24a) 2013, SpaceWeather, *What is a Coronal Hole?* http://www.spaceweatherlive.com/en/help/what-is-a-coronal-hole

[25] 1967, Russell, C.T. and Luhmann, J.G. from *Mercury: Magnetic Field and Magnetosphere* pages 476-478 originally published in the Encyclopedia of Planetary Sciences http://www-ssc.igpp.ucla.edu/personnel/russell/papers/merc_mag/

[26] 2012, April 5, The European Space Agency, *A Magnetic Surprise for Venus Express.* http://sci.esa.int/venus-express/50246-a-magnetic-surprise-for-venus-express/

[27] 2010, April 14, National Geographic Daily News, *Mini Magnetic Shield Found On the Moon* http://news.nationalgeographic.com/news/2010/04/100414-moon-magnetosphere-solar-wind/

[28] 2010, *Martin Wieser, a senior scientist at the Swedish Institute of Space Physics*, article by Anne Minard, The National Geographic, Daily News. http://news.nationalgeographic.com/news/2010/04/100414-moon-magnetosphere-solar-wind/

[29] 2013, November 12, NASA Science News, quoting Bruce Jakosky, Principal Investigator for MAVEN at the University of

Colorado at Boulder. http://science.nasa.gov/science-news/science-at-nasa/2013/12nov_maven/

[30] 2000, December 15, A University of California at Berkeley, Press Release by Robert Sanders.

[31] 2008, Seymour, Percy *Dark Matters: Unifying Matter, Dark Matter, Dark Energy, and the Universal Grid* (L474) Kindle Edition.

[32] 2011, February 7, NASA, *Jupiter's Magnetosphere: The Largest in the Solar System*, http://solarsystem.nasa.gov/scitech/display.cfm?ST_ID=1589

[33] 2004, NASA-JPL, *About Saturn and It's Moons* http://saturn.jpl.nasa.gov/science/index.cfm

[34] 2008, NASA, *Uranus*, http://solarsystem.nasa.gov/planets/profile.cfm?Object=Uranus&Display=OverviewLong

[35] 2013, Choi, Charles Q. Space.com, *How the Mighty Winds of Uranus and Neptune Blow,* http://www.space.com/21157-uranus-neptune-winds-revealed.html *Researchers Find Winds on Uranus Confined to Thin Atmospheric Layer,* http://phys.org/news/2013-05-uranus-neptune-confined-thin-atmosphere.html SolStation, Neptune Breaking News, http://www.solstation.com/stars/neptune.htm

[36] 2013, Windows to the Universe, *A Look at Pluto's Possible Magnetosphere*, http://www.windows2universe.org/pluto/magnetosphere.html

[37] 2011-2013, University of Berkley, *Themis* http://cse.ssl.berkeley.edu/artemis/mission-aurora-explain.html

[38] 2013, NASA, *Noctilucent Clouds Get An Early Start*, http://science.nasa.gov/science-news/science-at-nasa/2013/07jun_nlcs/ *Electric-blue Clouds Appear Over Antarctica.* http://www.youtube.com/watch?v=GHpxD807kM4

[39] 2006, April 23, NASA, *The History of Auroral Substorms.* http://www.nasa.gov/mission_pages/themis/auroras/substorm_history.html

[40] 2012, 2013, Encyclopedia Britannica, *Ionosphere and Magnetosphere*, http://www.britannica.com/EBchecked/topic/1369043/ionosphere-and-magnetosphere
NASA, Van Allen Probes: http://www.nasa.gov/mission_pages/rbsp/main/index.html

[41] 2009, Akasofu, Syun-Ichi *The Northern Lights* page 107,

Alaska Northwest Books
[42] 2013, Real time, instant space data:
http://www.solarham.net/data.htm
[43] 2013, March 8, NASA,
Ion Cyclotron Waves: Solar Wind Energy Source Discovered
http://science.nasa.gov/science-news/science-at-
nasa/2013/08mar_solarwind/
[44] 1999, NASA, *The Source of the Solar Wind Discovered*,
http://sohowww.nascom.nasa.gov/hotshots/1999_01_03/
[45] 2002, NASA, *Interplanetary Magnetic Field Lines*,
http://www-spof.gsfc.nasa.gov/stargaze/Simfproj.htm
[46] 2013, Chinese Academy of Science, *Earth's Magnetopause and Bow Shock* http://eng.sepc.ac.cn/MBS.php
[47] 2001, NASA, *The Tail of the Magnetosphere*,
http://www-spof.gsfc.nasa.gov/Education/wtail.html

Space Transference

Between each of the planets is an impressionable medium affected by the weight of passing spheres of energy. The weight of energy upon this impressionable medium creates magnetic waves. [48]

Professor Richard P. Feynman, winner of the Nobel Prize for Physics (1965) said, "If, in some cataclysm, all of scientific knowledge were to be destroyed and only one sentence passed on to the next generation of creatures, what statement would contain the most information in the fewest words? I believe it is the atomic hypothesis that: All things are made of atoms - little particles that move around in perpetual motion, attracting each other when they are a little distant apart but repelling upon being squeezed into one another. In that one sentence, you will see, there is an enormous amount of information about the world, if just a little imagination and thinking are applied." [49]

I propose that these atoms escape from every planet in our solar system by leaving through the auroral ovals to enter interplanetary space where energy push gives way to magnetic pull.

American Astronomer, Professor Henry Norris Russell, of Princeton, said in the Scientific American for November 1929, when talking about Einstein's contribution. "The central fact which has been proved— and which is of great interest and importance— is that the natural phenomena involving gravitation and inertia (such as the motions of the planets) and the phenomena involving electricity and magnetism (including the motion of light) are not independent of one another, but are intimately related, so that both sets of phenomena should be regarded as parts of one vast system, embracing all Nature. The relation of the two is, however, of such a character that it is perceptible only in a very few instances, and then only to refined observations." [50]

Sources for Space Transference

[48] 2012, Rhythmodynamics – *Matter is Made of Waves*: http://www.rhythmodynamics.com/Gabriel_LaFreniere/
[49] 1965, Feynman, Richard, *Six Easy Pieces* page 4, http://www.nobelprize.org/nobel_prizes/physics/laureates/1965/feynman-lecture.html
[50] 1929, November, Scientific American.

Planetary Dispersion

On Earth, we have a fairly constant and consistent large scale structure that facilitates atmospheric circulation, the large scale movement of air. The wind belts surrounding our planet are broken into three longitudinal circulation cells. [51]

1. Hadley Cell (*cardinal*): The Hadley Cell is a high volume closed circulation loop, where the vast bulk of vertical motion that occurs is similar in motion to the astrological *cardinal* activity. Moist air is lifted aloft in the inter-tropical convergence zone to the troposphere and carried to the poles.

2. The Polar Cell (*fixed*): The Polar Cell is a stable area that strengthens and weakens, driven by lower latitude warm air and undergoes enough convection to drive a thermal loop. The Polar Cell is similar in motion to the astrological *fixed* activity. The cold air in the polar areas mixes with the auroral flow and drops as cold dry high pressure moving away from the poles to travel westward. This downward outflow creates harmonic waves in the atmosphere know as Rossby waves.

3. Ferrel Cell (*mutable*): The Ferrel Cell is similar in motion to the astrological *mutable* activity as it is the circulation facilitator for the high and low pressure areas. This cell is sometimes known as the "zone of mixing." The Ferrel Cell is not a closed loop like the Hedley and Polar Cells where the flows are more predictable. The Ferrel Cell is the movement of air masses influenced by the jet stream that I propose has been influenced by an auroral oval flow.

Rossby waves, (often called planetary waves) are very long high frequency westward moving waves that guide the path of the jet stream traveling the troposphere and enter the Ferrel Cell. Rossby waves are found both in the atmosphere and in the oceans. [52]

Between these three cells is an overall atmospheric motion called zonal overturning circulation, [53] a consequence of solar radiation density that like an astrological *cusp* merges the cells together at their outer perimeters.

There are vast ocean currents like the Pacific Ocean Cell, an entirely ocean based circulation cell that flows below the atmospheric wind circulation cells, affecting our weather as energy is transferred from ocean to atmosphere. This is where El Nino and La Nina are born. [54]

In these days of extreme weather, life here on Earth is easily disrupted by energy from our atmosphere that has come through the auroral oval. Energy is transferred from our atmosphere to human life here on Earth. I propose that this energy should contain astrological signatures from all the planets in our solar system.

The planets are always dispersing and our planet Earth is receiving all the time. The cyclical nature of the planets may cause increases and decreases to the dispersing and receiving.

Can we go so far as to say that something transported through space from Saturn has reached Earth and is playing a role in our weather? Possibly, but we will only know for sure when science produces a cross section of the inflow of Earth's northern auroral oval. A cross section that will identify quality, quantity and the originating source.

Can we go further and say that something in our atmosphere that has arrived from a planet in our solar system is affecting our health? Possibly, if we are breathing our atmosphere then we are breathing in part what has come through the auroral oval, the gateway to the other planets of our solar system.

Magnetism and electricity are all around us. The human body is composed of a high quantity of water making it an excellent receptor and conductor of magnetic electricity energy. Science has recently discovered that magnetotactic bacteria contain inside them chains of very small single crystals of magnetite. These chains line up with the Earth's magnetic field and the bacteria themselves are then also aligned, unable to rotate away from this fixed direction, but confined to travel up and down a magnetic field line. [55] Growth of such mineral structures inside living organisms is called biomineralization. Examples of this process of mineral growth include the formation of shells and bone. [56]

Since I wrote the first draft of CPI Theory Part One, I have received comments and suggestions from several prominent astrologers. I have tried to incorporate all that I have received and clarify my position and thinking. Any errors are mine. Any incompleteness is my inability at this time to find scientific evidence. The purpose of this theory is to open future thought on the electromagnetic connections between planets and interplanetary space and the ramifications for life on Earth.

Sources for Planetary Dispersion

[51] 2013, *General Circulation of the Atmosphere,*
North Carolina State University
http://www.nc-climate.ncsu.edu/edu/k12/.atmosphere_circulation
[52] 2015, Columbia University, *Kelvin and Rossby Waves,*
http://iri.columbia.edu/climate/ENSO/theory/waves.html
[53] 2008, Science Daily, *Atmospheric Circulation,*
http://www.sciencedaily.com/articles/a/atmospheric_circulation.htm
[54] 2010, *Understanding Deep Ocean Circulation and Climate Modeling,* http://arstechnica.com/science/2010/01/understanding-deep-ocean-circulation-and-climate-modeling/ University of Wisconsin, *Ocean Circulation and Atmospheric Circulation,* http://www4.uwsp.edu/geo/faculty/ritter/geog101/textbook/circulation/ocean_circulation.html
University of Oregon, *The Hydrosphere an overview*
http://jersey.uoregon.edu/~mstrick/hydrosphere/hydro_overview.html
[55] 2012, Blundell, Stephen, *Magnetism: A Very Short Introduction,* Oxford University, Kindle Edition (L 1491)
[56] 2012, Blundell, Stephen, *Magnetism: A Very Short Introduction,* Oxford University Press, Kindle Edition (L 1495)

Summary

To summarize the Abstract

- *electron beams from interplanetary space hit Earth along a ring shaped oval where cathode rays interact with high speed electrons*

- *planetary discharges arrive channelled by the magnetic field*

- *biophysical processes are acted upon by human beings*

To summarize the Introduction

- *new thinking is needed to solve long-standing unsolved problems*

- *a clear statement of fact that leaves no doubt that planetary influence is alive and well*

- *each planet is surrounded by an interactive sphere*

- *an astrological signature from each of the planets should be entering Earth's atmosphere*

To summarize Planetary Sphere

- *spheres are quite common on Earth*

- *interaction happens at the auroral oval*

To summarize Auroral Oval

- *auroral oval acts as a generator producing up to 10,000,000 megawatts of electrical power*

- *on average 1,000,000 ampere current flows through the auroral curtain*

- *auroral activity and geomagnetic disturbances are only different manifestations of an enhanced dynamo power*

- *auroral light displays cause the Earth's atmosphere to expand slightly*

- *each planet through its auroral ovals will disburse or receive material*

- *each planet will have a unique signature*

- *ionized atoms and electrons flow along the open magnetic fields in coronal holes*

- *a mini magnetosphere opens the door for finding magnetospheres on smaller bodies, even asteroids*

- *magnetosphere distorted to the point that it is currently unrecognizable*

- *an electromagnetic disbursement and capture mechanism through auroral ovals*

- *Saturn's moon Enceladus has an auroral oval*

- *Earth's northern auroral oval has a continuous curtain around its outer edge*

- *the auroral curtain hangs down into Earth's atmosphere*

- *auroral light can at times be seen over vast areas*

- *plasmas with electrons traveling the field lines spiral in and hit the atmosphere*

- *when auroral activity increases, more galactic radio noise is absorbed into the ionosphere*

- *the continuous never stopping solar wind made up of electrically charged particles*

- *the almost constantly changing solar wind affects the voltage of the auroral generator*

- *Earth's magnetosphere is formed by the impact of the solar wind on the Earth's magnetic field*

- *the flow pattern of magnetospheric plasma moves from the magnetotail toward the Earth*

- *the magnetosphere is full of trapped plasma*

To summarize Space Transference

- *the weight of energy upon this impressionable medium creates magnetic waves*

- *all things are made of atoms*

- *atoms escape from every planet in our solar system*

- *interplanetary space where energy push gives way to magnetic pull*

To summarize Planetary Dispersion

- *the Hadley Cell is a high volume closed circulation loop*

- *the Ferrel Cell is similar in motion to the astrological mutable activity*

- *the Polar Cell is a stable area that strengthens and weakens*

- *Rossby waves are found both in the atmosphere and in the oceans*

- *an overall atmospheric motion called zonal overturning circulation*

- *energy is transferred from our atmosphere to human life here on Earth*

- *this energy should contain astrological signatures from all the planets in our solar system*

- *a cross section that identifies quality and quantity and originating source*

- *we are breathing in part what has come through the auroral oval*

- *magnetism and electricity are all around us*

- *chains of very small single crystals of magnetite line up with the Earth's magnetic field*

- *to open future thought on the electromagnetic connections between planets*

Conclusions for Part One

The interaction between an auroral oval and a solar system planet seems clear. The auroral oval receives and disburses electron energy. Between the planets in interplanetary space is a medium that is electromagnetic and diverse.

Electromagnetism communicates with itself and can be found throughout the solar system.

Acknowledgements for Part One

I would like to thank the following Astrologers whose comments helped guide CPI Theory Part One to publication: Firstly in the USA, Bruce Scofield, PhD., who watered the original seeds and caused me to start writing; Robert Currey in the UK for his exceptionally strong and unending support; John Rutherford in Canada, who tried to keep me on a scientific track; Robin Armstrong in Canada, who gave early encouragement; Chand Karan Ahuja, B.Sc in India, who brought a tear to my eye; Michael Erlewine in the USA, who reminded me that it is all about people; Jagdish C. Maheshri, Ph.D., in the USA, for his kindness in helping prepare for publication. And, thanks to all who offered support to my efforts. Any mistakes are mine.

PART TWO

Abstract

In CPI Theory Part One first published as the Theory of Continuous Planetary Interaction in the 2014 NCGR Research Journal, a case was made that the auroral oval is a key mechanism that allows a solar system planet to electromagnetically interact with interplanetary space and with another planet in our solar system. Within auroral ovals the ionosphere occupies a large space that facilitates the exchange of electron encounters. This leads to changes and modifications of atoms and molecules. Taking a larger picture of our solar system it can be seen that the electron is a universal quality essential to the electromagnetic movement and exchange of energy between planets.

CPI Theory Part Two suggests that electron encounters and astrological planetary energy transference may occur at the same time and place.

Introduction

CPI Theory Part Two explores:

(1) The auroral oval as an integral mechanism of continuous planetary interaction.
(2) The creative role of the electron within our solar system.
(3) The important role of the ionosphere within the auroral oval.
(4) How interplanetary space interacts with the auroral oval and the ionosphere.

The auroral oval is one process of several that link the planets together but it is a very important process that leads to atomic reactions and modifications that produce an effect on living matter here on Earth.

In CPI Theory Part One it was stated that Earth, Mercury, Jupiter, Saturn, Uranus, Neptune and Pluto have auroral ovals. The Sun has coronal holes that may prove to be a variation on the auroral oval. Venus and Mars each have ionospheres that are a key component of auroral oval activity while they each have cases building for auroral ovals that are yet to be identified. The Moon is the hardest to show auroral oval activity but the more that magnetic sources continue to be found on the Moon the stronger the likelihood of a magnetic field that leads to an auroral oval. The Moon does have an ionosphere that extends up 5 kilometers from the surface. Twilight rays were witnessed on the Moon by crewmembers of Apollo 8, 10, 15 and 17. [58]

In CPI Theory Part One the auroral oval was shown as a mechanism that facilitated transfers between ionized interplanetary space and matter within a planet's electromagnetic atmosphere. But what is interacting and how?

Now CPI Theory Part Two adds another mechanism to the case for continuous planetary interaction.

CPI Theory Part Two will discuss how the electron is a universal quality that becomes the fundamental source of interaction from interplanetary space to the surface of a planet to interact with all life. Electrons are an integral part of the atomic nucleus facilitating the formation of atoms and matter. [59]

The auroral oval is found at the poles of a magnetosphere surrounding a planet where charged particles inflow from interplanetary space. Charged particles interact with the Earth's magnetic field and spiral around the magnetic field lines, concentrating at the poles. Due to the high density of the magnetic field lines collisions occur between particles, resulting in emissions of light. [60]

Magnetic fields produce a force on charged particles causing the charged particles to execute corkscrew paths which wind around field lines. Many charged particles from the solar wind become trapped in doughnut-shaped regions around the Earth, known as the Van Allen belts. [61]

The auroral oval disburses planetary outflow via magnetic field lines out into interplanetary space.

The planets of our solar system are surrounded by spheres such as a magnetosphere, an ionosphere or an atmosphere. Within these spheres are various layers of temperature. The auroral oval accepts electrons from interplanetary space that then enter a planetary sphere. Here on Earth magnetic interaction causes the traveling electron to interact with the contents of each thermal layer on its path through the ionosphere and into the planetary atmosphere.

What follows will be in three chapters:

The Electron
The Auroral Oval
Planets and Interplanetary Space

Before moving ahead a word on the electron is necessary. The electron is extremely complex and diverse. The electron has some very remarkable qualities that can also be expressed in very simple terms. The electron has been given so many names by science along with confusing units that its continuity is hard to see in full perspective even though most of what is going on around and within us is electrons interacting in one form or another. [62]

Electrons are found on all planets in our solar system. Electrons pervade interplanetary space that surrounds the planets. Electrons are a self-energizing transportation common quality. Electrons are a fundamental component for the assembly and disassembly of atoms within an electromagnetically conducting ionosphere especially when energized by a magnetosphere with an auroral oval.

Sources for the Introduction

[58] 2013, NASA; http://science.nasa.gov/science-news/science-at-nasa/2013/03sep_ladee/
[59] 2011, Hall, Alan *Electron* (L 301) Kindle Edition.
[60] 2012, Blundell, Stephen J. *Magnetism: A Very Short Introduction* Oxford University Press, (L 1614) (p. 111) Kindle Edition.

[61] 2012, Blundell, Stephen J. *Magnetism: A Very Short Introduction* Oxford University Press, (L 1618) (p. 111) Kindle Edition.
[62] 2008, Harvard University http://hea-www.harvard.edu/astrostat/slog/groundtruth.info/AstroStat/slog/2008/eotw-kev-kev/index.html

The Electron

In 1897, J. J. Thomson at the Cavendish Laboratory in Cambridge, England discovered the electron, the first subatomic particle. [63] In 1906, J. J. Thomson won the Nobel Prize for showing that the electron is a particle. In 1937, George Thomson the son of J.J. Thompson won the Nobel Prize for showing that the electron is a wave. [64]

An electron appears to be truly fundamental. [65] It embraces gases, liquids and mass. Electrons are seen through the spectrum of light and identified as wave frequencies. Electrons determine the color, reactivity, and shape of chemicals. [66] The electron exists in the coldest astrological planet of our solar system Pluto (average temperature -380 Fahrenheit) and in the hottest astrological planet the Sun (average temperature 8 million degrees Fahrenheit). [67] The electron also exists at all temperatures within these extreme ranges. In 2006 the International Astronomical Union in a controversial unilateral decision decided that Pluto no longer qualifies as a planet. Most western astrology today includes Pluto as an astrological planet.

John Polkinghorne, Professor of Mathematical Physics says, "I believe it is because the assumption that there are electrons, with all the subtle quantum properties that go with them, makes intelligible great swathes of physical experience that otherwise would be opaque to us. It explains the conduction properties of metals, the chemical properties of atoms ... and much else besides." [68]

The central nucleus of an atom is surrounded by orbiting negatively charged electrons. [69] In 1909 Ernest Rutherford concluded that the atom was rather like a miniature solar system, with a very small and dense central nucleus with electrons orbiting around it, like the planets orbiting the Sun. [70] Electrons function to contain nuclei and thus form atoms, molecules and matter. [71]

Robin Armstrong, President of the RASA School of Astrology says, "Paralleling the solar system to an atom opens many profound ideas. DNA structures and diseases for example. The astrological implications of the solar system micro-associated to the atom and DNA structures could provide a quantum leap of subjective associations and subsequent relevance." [71a]

Dr. Alan Hall says, "... it is the humble electron that holds the secret of how atoms interact and bind together to form molecules ... It is the spin of the electron and its magnetic moment that generates the electromagnetic force." [72]

The electromagnetic force that holds electrons in atoms and links atoms to one another to make molecules holds matter together. [73] The atoms of the elements are more often than not joined into the unions called molecules. [74] These molecules have electromagnetic binding energies measured by the energies of the orbiting electrons holding them together. [75]

Chemistry is what happens when electrons stick atoms together to make molecules. The electrons doing the sticking are the outer ones, those furthest from the nucleus. [76] Electrons that are the farthest from the nucleus will determine the chemical interactions of the atom with other atoms. [77]

Atoms can combine, or react, in many different ways. A chemical reaction means that the electrons holding atoms together are rearranging themselves. [78] Atoms bind together by sharing electrons to make molecular structures, but these bonds can be broken. Fossil fuels release energy when intermolecular bonds break. [79] Burning fuel in a fire, is extracting electromagnetic energy by causing the electrons holding together the molecules in the fuel to change. [80] When food is digested we extract electromagnetic energy from the food by changing the electron orbits in its molecules. [81]

The electron is fundamental to atomic elements. Electrons have been inside atoms as long as the Earth has been here. [82] The number of electrons in an uncharged atom defines the chemical elements. Every hydrogen atom in the universe has one electron, every helium atom two electrons, carbon has six electrons, nitrogen seven and oxygen has eight. [83] The simplest atom of all, corresponding to the lightest element, hydrogen, has a nucleus consisting of a single proton, balancing the charge of the single orbiting electron. [84]

An electron in an atom can escape the electrical pull of the atomic nucleus. [85] Metals conduct electricity because some of their electrons come free of their parent atoms and are at liberty to roam through the material where their motion corresponds to an electrical current. They are not intrinsically free, but can be shaken loose from their atoms by mild heat. Some electrons are liberated at room temperature. [86]

At room temperature there will be slightly more, lower energy down spins than higher energy up spins. [87] The electron is able to influence the space which surrounds it by its spin. [88] On average, half of the electrons will spin one way and half will spin the other way. [89] The spinning motion of the electron gives rise to a magnetic field. [90] Electrons generate a magnetic moment because they spin. It is the direction of that spin that determines whether various electrons attract or repel each other. [91]

The electron appears to be releasing or creating energy continuously, possibly due to constantly aligning its gyroscopic spin direction to a plane that we have yet to name. It is known that an increasing magnetic field allows energy to escape as the electron spins realign. [92]

A team of researchers from the International School for Advanced Studies (SISSA) and International Centre for Theoretical Physics Abdus Salam of Trieste observed that the four oxygen molecules found in a quartet grouping were constantly exchanging magnetic moments. Professor Erio Tosatti from SISSA says, "It's as if the molecules were playing ball with their spins, the direction in which the electrons rotate around their axis, continuously passing the ball to one another, so that the mean value of each molecule's moment and magnetism is zero." [93]

As electrons in an atom encounter increasing temperature they are raised to ever higher energy levels, until eventually they are ejected from the atom leaving the atom ionized. [94] An ionized atom has a positive or negative electrical charge and an ion is an atom with a positive or negative electrical charge. [95] Metals form positive ions by the loss of electrons, while non-metals gain electrons to form negative ions. [96]

The electron as far as we know is itself fundamental. [97] One of the key motivators of the electron is temperature. It is the exterior temperature at the electron's location that facilitates the type of energy the electron displays. Electrons can simultaneously carry electrical charge and conduct heat.

Electrons move in response to either an electric field or a temperature gradient. [98]

Temperature quantifies the degree to which a system can vibrate and fluctuate by exchanging energy with its environment; warmer molecules move around more than cold ones. At a certain temperature, −273.15 Celsius (−459.67 Fahrenheit) also known as absolute zero, all vibrations cease. [99] Most bodies broadcast energy in a wide range of wavelengths, from X-rays to very long radio waves. At a given temperature, most of the radiation given off will be close to a specific wavelength creating a specific relationship between the wavelength at which most of the energy is emitted, and the temperature. [100]

It is known that an electric current is made up of charged particles. [101] When ever electrons are disturbed we see the advent of electromagnetism. [102] An orderly flow of electrons produces an electrical current. [103]

According to quantum mechanics, electrons behave as much as waves as they do as particles. The 1933 Nobel Prize for Physics went to Professor Erwin Schrödinger whose equation identifies the possible quantum states that the electrons in an atom can find itself. In 1966 Physicists Professor Friedrich Hund and Professor Robert S. Mulliken both won a Nobel Prize in Chemistry. Each had independently developed the molecular orbital theory in which bonding occurred as a result of constructive and destructive interference between electron waves on each atom in a molecule. [104]

The electron orbits a particle and through the electromagnetic currents produced by the electron's orbiting and internal spins, the electron excites an atomic nucleus to life. One orbiting electron can create a simple atom or can sequentially combine as pairs to create complex atoms with many electrons orbiting the atomic nucleus in conflict free orbits. Stephen J. Blundell, Professor of Physics says, "Quantum mechanics forces the orbits of electrons around the atom to take certain fixed configurations with certain allowed speeds of rotation." [105]

Two clockwise rotating electrons will be magnetically attracted to each other while opposite rotating electrons will be mutually repelled. [106] The electrical interactions between the electrons of one chemical element and those of another chemical element are responsible for the combining of elements to form chemical compounds. The forces that combine all solid materials together are also associated with the behavior of the electrons. [107]

Quantum theory shows that the electrons orbits are not random, but are fixed in a small set of allowed orbits, each one of which is associated with a fixed value of energy. Light can be emitted when an electron transfers from one orbit to another. The energy of the emitted light makes up for the difference between the energies in the two orbits. [108]

The electron does not always possess a definite position or a definite momentum, but does possess the potentiality for exhibiting one or the other. [109] Electrons are so small that probabilities are used to describe where they are likely to be found. [110] The electron's probability is spread out, in every direction. [111]

Professor John Polkinghorne says, "It is possible to know where an electron is, but not know what it is doing and it is possible to know what it is doing, but not know where it is." [112]

The mass of an electron can be determined by the extent to which it is deflected in a magnetic field. [113] The mass of the electron helps determine the size of atoms and is in part responsible for things being the size they are. [114]

It is a fundamental law of nature that mass and energy are just two forms of the same thing. [115] Paul Dirac, Theoretical Physicist and a founder of quantum physics described the energy of the electron as the sum of its rest mass energy and the contribution of its momentum. [116] The mass of an electron moving close to the speed of light would be much more massive than an electron moving at a much slower speed. [117]

Eric R. Scerri, Professor of Chemistry says, "The need to take account of relativity arises whenever objects move at speeds close to that of light. Inner electrons, especially those in the heavier atoms in the periodic system, can readily attain such relativistic velocities ... In addition, many seemingly mundane properties of elements, like the characteristic colour of gold or the liquidity of mercury, can best be explained as relativistic effects due to fast-moving inner-shell electrons." [118]

Materials can be classified in terms of how well they conduct electricity. Metals, like copper and gold, are good conductors of electricity and are used for making wires. The electrons in a metal are able to easily travel around. Many (though not all) plastics and rubbers tend to be poor conductors of electricity, and are classified as insulators. The electrons in an insulator are fixed in place and are not able to travel around easily. [119]

Science through its various disciplines offers us a zoom lens to peer deeper into the small or to scope out into large perspectives but at all times the conversation centers around the electron. If there is no electron the atom remains unexcited and unrealized. Spectroscopy has shown that just like planet Earth, the electron has a magnetic field and that it has magnetic poles with a north and south polarity. [120]

Professor Stephen J. Blundell says, "...electrons exist in two different spin states and so a macroscopic electric current contains electrons of both spins. That means that the spin-up electrons can travel much more easily and travel a path of low resistance. For spin-down electrons, it's much more like they are wading through treacle." [121] I propose that this difference in resistance between spin-up and spin-down could be similar to that of apparent motion of the planets orbiting around the Sun in astrology.

In an atom, magnetism arises from the spin and orbital momentum of its electrons. The electrons move in two ways: (1) spin, which can loosely be thought as spinning around themselves; (2) orbit, which refers to an electron's movement around the nucleus of its atom. The spin and orbital motion gives rise to the magnetization producing a magnetic field. The spinning direction of the electrons therefore defines the direction of the magnetization in a material. [122]

The periodic table has led to a deep knowledge of the structure of the atom and the notion that electrons essentially circle the nucleus in specific shells and orbitals. [123] The most recent explanation of the periodic system is given in terms of how many orbitals are populated by electrons. The explanation depends on the electron arrangement, or configuration, of an atom, which is spelled out in terms of the occupation of its orbitals. [124]

The electron may remain in orbit around a particular atom or it may set itself free to singularly orbit another particle. Or it can combine with other electrons thus inducing an atomic nucleus to become a different atomic element of a certain nuclear character. These are enumerated in the periodic table where the structure of the periodic table reflects the quantum numbers of the orbitals that hold electrons. [125] The 92 elements that make up all the matter we find in nature differ from each other in the number of electrons they have ranging from 1 to 92. [126]

The blending of quantum mechanics with the periodic table created an understanding based upon electronic configurations. In this approach, the elements in the periodic table differ from each other according to the type of orbital occupied by the last electron to enter the atom in the building-up process. [127]

Electrons reside in very specific orbitals, and these orbitals have very specific energies. [128] Electron orbitals possess nodes. [129] A node is a place where there is zero probability of finding an electron. [130] The more nodes within an orbital, the higher the energy of that orbital. [131]

After evolving for nearly 150 years the periodic table remains at the heart of the study of chemistry. This is mainly because it is of immense practical benefit for making predictions about all manner of chemical and physical properties of the elements and possibilities for bond formation. [132]

Physicist, Antonius Van den Broek suggested that since the nuclear charge on an atom was half of its atomic weight, and that the atomic weights of successive elements increased in step-wise fashion by two, then the nuclear charge defined the position of an element in the periodic table. In other words, each successive element in the periodic table would have a nuclear charge greater by one than the previous element. [133] The larger the nuclear charge the faster the motion of inner shell electrons. [134]

The physical rules governing the arrangement of electrons around the nucleus mean that atoms divide into families characterized by their outer electron configurations. Since the outer electrons specify the chemical properties of the elements, these families have similar chemistry. [135]

Quantum theory was actually born in the year 1900. It was first applied to atoms by Niels Bohr, who pursued the notion that the similarities between the elements in any group of the periodic table could be explained by their having equal numbers of outer-shell electrons. Electrons are assumed to possess only certain quanta, or packets, of energy and, depending on how many such quanta they possess, they lie in one or another shell around the nucleus of the atom. [136] Chemical behavior of an element depends on the quantity and arrangement of its electrons in their shell structure. [137]

In the case of the hydrogen atom Niels Bohr was able to show that its single electron could, according to quantum principle, only orbit at certain fixed distances from the nucleus. When the electron was in one of these fixed orbits it would not radiate, but when it jumped from one orbit to the next, it would either radiate some energy or absorb some energy. [138]

It is the order of filling rather than the distance of electrons in various types of orbitals from the nucleus that should be regarded as being more fundamental. [139] Electron shells begin by filling in a sequential manner. There is an assumption that electron shells are filled sequentially as one moves through the periodic table adding an extra electron for each new element encountered. [140] This is only true for the first 18 elements of the periodic table and ceases to be the case starting with potassium, element number 19. [141]

When the atom absorbs electromagnetic waves, such as light, the electron is shifted from an orbit close to the nucleus to one further away. Later, the electron will fall from its excited state broadcasting light at a precise wavelength. Moving from an orbit far from the nucleus to one closer to the nucleus will bring about a different wavelength, so the atom will emit a well-defined series of wavelengths characteristic of that particular type of atom. [142]

It was Maxwell who formulated the full theory of electromagnetism, which unified electricity, magnetism, and light. As new types of radiation were discovered, X-rays, radio waves, infrared, and ultraviolet light, it was realized that they too were part of what is now known as the electromagnetic spectrum. [143] X-ray frequency emissions occur because an inner electron is ejected from the atom, causing an outer electron to fill the empty space, in a process which is accompanied by the emission of X-rays. [144]

An electron can become a cell membrane and life. Everything that lives on the surface of the planet is cellular in nature. [145] All living organisms are complicated dissipative structures. [146] From the interior of the nucleus to the surface of the cell, there are links between just about all of the filamentous proteins. [147]

All cells reproduce themselves by splitting into two. Some bacteria manage to increase their contents rapidly enough to undergo division by a process known as binary fission in as little as 20 minutes. The much larger eukaryote cells may take the best part of a day to double their size before division. [148] The dynamic nature of the cell membrane is such that an entire plasma membrane is replaced on an hourly basis. [149] There are extreme levels of synthesis that allow the simpler cells to double themselves in minutes, and more complicated cells within a day. At the fundamental level, life is based on the atoms of only six of the 117 {118 in 2015} known elements: carbon, hydrogen, nitrogen, oxygen, phosphorus, and sulphur. [150]

Life generates energy from microscopic electrical motors that are embedded in cell membranes and run off electrical currents driven by pH gradients across the membranes. When discharged, electrical current flows through the molecular turbine in the cell membrane and generates molecules that store energy. [151]

Dr. Graham Cowling and Terence Allen, Honorary Professor of Structural Cell Biology say, "The cell is the basic unit of life, and as such must fulfill three requirements: (1) to be a separate entity, requiring a surface membrane; (2) to interact with the surrounding environment to extract energy in some way for maintenance and growth; and (3) to replicate itself. These parameters are the same for all living beings, from the smallest bacterium, to any one of the 200 different cell types that create a human being." [152]

A cell can function perfectly well as a single entity or, alternatively, one cell may be an infinitesimally small part of a massive community of cells that work together to make a single being such as ourselves. [153] Every cell has a blueprint for its own creation coded by the DNA of its genes. In any particular organism, the DNA information content is the same in every cell type, whether they are brain or skin cells. [154]

Cell shape is controlled from within, modulated by signals received from the external environment, and capable of rapid response. Cells continually change their shape, change position relative to their neighbours, move through solid tissues, or take long journeys around the body by entering and exiting the bloodstream. [155] It is the electrons in the atoms of these cells that control what is happening. [156]

In the superconducting state, an enormous number of electrons act in concert, as if each is part of a larger, inseparable whole. [157] The theory of superconductivity shows how electrons pair up and how these pairs can together form a giant collective unit in which the phases of all the individual wavefunctions lock together. [158]

In superconductivity, we see the quantum-mechanical ganging up of the electrons in their many-body paired state, in which the individuality of electron pairs is sacrificed to the greater good of collective unity. Superconductivity cannot occur with one atom. Superconductivity only occurs in an assembly of atoms, in much the same way that an orchestral symphony can only be played by an assembly of musicians. [159]

Andrew King, Professor of Astrophysics says, "Although gravity and electromagnetism correctly describe all the physics of stars on any length scale bigger than an atom, we know that these two forces are unable to provide the energy source that keeps the stars shining. So to find this source, we must look inwards, to physics on subatomic scales." [160]

Sources for The Electron

[63] 2011, Scerri, Eric R. *The Periodic Table: A Very Short Introduction* Oxford University Press, (L 1229) (p. 73) Kindle Edition.
[64] 2002, Polkinghorne, John. *Quantum Theory: A Very Short Introduction* Oxford University Press, (L 472) Kindle Edition.
[65] 2004, Close, Frank. *Particle Physics: A Very Short Introduction* Oxford University Press, (L 284) Kindle Edition.
[66] 2013, Lear, Benjamin; Lear, Shana. *Electron Orbitals and Electron Configurations* (L 276) Kindle Edition.
[67] 2016, Space.com; http://www.space.com/18563-pluto-temperature.html
[68] 2002, Polkinghorne, John. *Quantum Theory: A Very Short Introduction* Oxford University Press. (L 1395) Kindle Edition.
[69] 2011, Scerri, Eric R. *The Periodic Table: A Very Short Introduction* Oxford University Press, (L 1247) Kindle Edition.
[70] 2008, Seymour, Percy *Dark Matters: Unifying Matter, Dark Matter, Dark Energy, and the Universal Grid* (L54) Kindle Edition
[71] 2011, Hall, Alan *Electron* (L 301) Kindle Edition.
[71a] 2015, Armstrong, Robin, President of the RASA School of Astrology by permission November 2015 http://www.rasa.ws/
[72] 2011, Hall, Alan *Electron* (L 88) Kindle Edition.
[73] 2014, Close, Frank. *Particle Physics: A Very Short Introduction* Oxford University Press. (L 319) Kindle Edition.
[74] 2004, Ball, Philip. (L 229) *The Elements: A Very Short Introduction* Oxford University Press. Kindle Edition.
[75] 2012, King, Andrew. *Stars: A Very Short Introduction* (L 594) Oxford University Press. Kindle Edition.
[76] 2012, King, Andrew. *Stars: A Very Short Introduction* (L 590) Oxford University Press. Kindle Edition.
[77] 2002, Polkinghorne, John. *Quantum Theory: A Very Short Introduction* Oxford University Press. (L 1057) Kindle Edition.
[78] 2012, King, Andrew. *Stars: A Very Short Introduction* (L 594) Oxford University Press. Kindle Edition.
[79] 2012, Yee, Jeff. *The Particles of the Universe* (L 47) Kindle
[80] 2012, King, Andrew, *Stars: A Very Short Introduction* (L 596) Oxford University Press. Kindle Edition.
[81] 2012, King, Andrew. *Stars: A Very Short Introduction* (L 599)

Oxford University Press. Kindle Edition.

[82] 2004, Close, Frank. *Particle Physics: A Very Short Introduction* Oxford University Press. Kindle Edition. (L 601)

[83] 2012, King, Andrew. *Stars: A Very Short Introduction* (L 567) Oxford University Press. Kindle Edition.

[84] 2012, King, Andrew. *Stars: A Very Short Introduction* (L 618) Oxford University Press. Kindle Edition.

[85] 2012, Close, Frank. *Particle Physics: A Very Short Introduction* Oxford University Press. (L 463) Kindle Edition.

[86] 2004, Ball, Philip. *The Elements: A Very Short Introduction* Oxford University Press. (L 2075) Kindle Edition.

[87] 2010, Atkins, Peter *The Laws of Thermodynamics: A Very Short Introduction* Oxford University Press Kindle Edition. (L1447)

[88] 2011, Hall, Alan *Electron* (L 264), Kindle Edition.

[89] 2012, Blundell, Stephen J. *Magnetism: A Very Short Introduction* Oxford University Press. (L 1205) Kindle Edition.

[90] 2010, Atkins, Peter *The Laws of Thermodynamics: A Very Short Introduction* Oxford University Press Kindle Edition. (L 1447)

[91] 2011, Hall, Alan *Electron* (L 281) Kindle Edition.

[92] 2010, Atkins, Peter *The Laws of Thermodynamics: A Very Short Introduction* Oxford University Press Kindle Edition. (L 1458)

[93] 2014, Sissa Medialab via ScienceDaily.com http://www.sciencedaily.com/releases/2014/07/140707152312.htm?utm_source=feedburner&utm_medium=email&utm_campaign=Feed%3A+sciencedaily%2Ftop_news%2Ftop_science+%28ScienceDaily%3A+Top+Science+News%29

[94] 2004, Close, Frank. *Particle Physics: A Very Short Introduction* Oxford University Press. (L 688) Kindle Edition.

[95] 2013, Lear, Benjamin; Lear, Shana. *The Atom (A Brief Introduction to General Chemistry)* (L 64) Kindle Edition.

[96] 2011, Scerri, Eric R. *The Periodic Table: A Very Short Introduction* Oxford University Press. (L 618) (p. 20) Kindle Edition.

[97] 2004, Close, Frank. *Particle Physics: A Very Short Introduction* Oxford University Press. (L 754) Kindle Edition.

[98] 2014, Massachusetts Institute of Technology via ScienceDaily.com http://www.sciencedaily.com/releases/2014/07/140728104714.htm?

utm_source=feedburner&utm_medium=email&utm_campaign=Feed
%3A+sciencedaily%2Ftop_news%2Ftop_science+%28ScienceDaily
%3A+Top+Science+News%29

[99] 2009, Blundell, Stephen J. *Superconductivity: A Very Short Introduction* Oxford University Press. Kindle Edition. (L 377)

[100] 2008, Seymour, Percy *Dark Matters: Unifying Matter, Dark Matter, Dark Energy, and the Universal Grid* L 1642) Kindle Edition.

[101] 2008, Seymour, Percy *Dark Matters: Unifying Matter, Dark Matter, Dark Energy, and the Universal Grid* (L 1144) Kindle Edition.

[102] 2011, Hall, Alan *Electron* (L 301) Kindle Edition.

[103] 2010, Atkins, Peter *The Laws of Thermodynamics: A Very Short Introduction* Oxford University Press Kindle Edition. (L1341)

[104] 2011, Scerri, Eric R. *The Periodic Table: A Very Short Introduction* Oxford University Press. (L 1548) (p. 99) Kindle Edition.

[105] 2012, Blundell, Stephen J. *Magnetism: A Very Short Introduction*
Oxford University Press. (L 1120) (p. 73) Kindle Edition.

[106] 2011, Hall, Alan *Electron* (L 301) Kindle Edition.

[107] 2008, Seymour, Percy *Dark Matters: Unifying Matter, Dark Matter, Dark Energy, and the Universal Grid* (L 96) Kindle Edition.

[108] 2012, Blundell, Stephen J. *Magnetism: A Very Short Introduction*
Oxford University Press. (L 1107) (p. 72) Kindle Edition.

[109] 2002, Polkinghorne, John, *Quantum Theory: A Very Short Introduction* Oxford University Press. (L 1390) Kindle Edition.

[110] 2013, Lear, Benjamin; Lear, Shana. *Electron Orbitals and Electron Configurations* (L 4-5) Kindle Edition.

[111] 2002, Polkinghorne, John. *Quantum Theory: A Very Short Introduction* Oxford University Press. (L 817) Kindle Edition.

[112] 2002, Polkinghorne, John. *Quantum Theory: A Very Short Introduction* Oxford University Press. (L 653) Kindle Edition.

[113] 2008, Seymour, Percy *Dark Matters: Unifying Matter, Dark Matter, Dark Energy, and the Universal Grid* (L 1570) Kindle Edition.

[114] 2004, Close, Frank, *Particle Physics: A Very Short Introduction* Oxford University Press. (L 1641) Kindle Edition.

[115] 2012, King, Andrew. *Stars: A Very Short Introduction*
Oxford University Press. (L 636) Kindle Edition.

[116] 2011, Hall, Alan *Electron* (L 250) Kindle Edition.

[117] 2008, Seymour, Percy *Dark Matters: Unifying Matter, Dark Matter, Dark Energy, and the Universal Grid* (L1570) Kindle Edition.

[118] 2011, Scerri, Eric R. *The Periodic Table: A Very Short Introduction* Oxford University Press. (L 707) Kindle Edition.

[119] 2009, Blundell, Stephen J. *Superconductivity: A Very Short Introduction* Oxford University Press. (L 323) Kindle Edition.

[120] 2011, Hall, Alan *Electron* (L 281) Kindle Edition.

[121] 2012, Blundell, Stephen J. *Magnetism: A Very Short Introduction* Oxford University Press. (L 1360) Kindle Edition.

[122] 2014, Ecole Polytechnique Fédérale de Lausanne via ScienceDaily.com
http://www.sciencedaily.com/releases/2014/05/140508141829.htm?utm_source=feedburner&utm_medium=email&utm_campaign=Feed%3A+sciencedaily%2Ftop_news%2Ftop_science+%28ScienceDaily%3A+Top+Science+News%29

[123] 2011, Scerri, Eric R. *The Periodic Table: A Very Short Introduction* Oxford University Press. Kindle Edition. (L 373) (p. xix)

[124] 2011, Scerri, Eric R. *The Periodic Table: A Very Short Introduction* (L 716) Oxford University Press. Kindle Edition.

[125] 2013, Lear, Benjamin; Lear, Shana. *Electron Orbitals and Electron Configurations* (L 114) Kindle Edition.

[126] 2008, Stannard, Russell *Relativity: A Very Short Introduction* Oxford University Press. Kindle Edition. (L 683) (p. 39)

[127] 2011, Scerri, Eric R. *The Periodic Table: A Very Short Introduction* Oxford University Press. Kindle Edition. (L 1905) (p. 129)

[128] 2013, Lear, Benjamin; Lear, Shana, *Electron Orbitals and Electron Configurations* (L 15) Kindle Edition.

[129] 2013, Lear, Benjamin; Lear, Shana, *Electron Orbitals and Electron Configurations* (L 28) Kindle Edition.

[130] 2010, http://photonicswiki.org/index.php?title=Atomic_Orbitals_and_Nodes

[131] 2013, Lear, Benjamin; Lear, Shana, *Electron Orbitals and Electron Configurations* (L 53) Kindle Edition.

[132] 2011, Scerri, Eric R. *The Periodic Table: A Very Short Introduction* (L 726) Oxford University Press. Kindle Edition.

[133] 2011, Scerri, Eric R. *The Periodic Table: A Very Short Introduction* Oxford University Press. Kindle Edition. (L 1302)
[134] 2011, Scerri, Eric R. *The Periodic Table: A Very Short Introduction* Oxford University Press. Kindle Edition. (L 1794)
[135] 2012, King, Andrew, *Stars: A Very Short Introduction* (L 590) Oxford University Press. Kindle Edition.
[136] 2011, Scerri, Eric R. *The Periodic Table: A Very Short Introduction* Oxford University Press. Kindle Edition. (L 708)
[137] 2004, Ball, Philip, *The Elements: A Very Short Introduction* (L 1816) Oxford University Press. Kindle Edition.
[138] 2008, Seymour, Percy *Dark Matters: Unifying Matter, Dark Matter, Dark Energy, and the Universal Grid* (L 64) Kindle Edition.
[139] 2011, Scerri, Eric R. *The Periodic Table: A Very Short Introduction* Oxford University Press. Kindle Edition. (L 1920)
[140] 2011, Scerri, Eric R. *The Periodic Table: A Very Short Introduction* Oxford University Press. Kindle Edition. (L 1522)
[141] 2011, Scerri, Eric R. *The Periodic Table: A Very Short Introduction* Oxford University Press. Kindle Edition. (L 1445)
[142] 2008, Seymour, Percy *Dark Matters: Unifying Matter, Dark Matter, Dark Energy, and the Universal Grid* (L314) Kindle Edition.
[143] 2008, Seymour, Percy *Dark Matters: Unifying Matter, Dark Matter, Dark Energy, and the Universal Grid* (L212) Kindle Edition.
[144] 2011, Scerri, Eric R. *The Periodic Table: A Very Short Introduction* Oxford University Press. Kindle Edition (L 1316)
[145] 2011, Allen, Terence; Cowling, Graham, *The Cell: A Very Short Introduction* Oxford University Press. Kindle Edition. (L 304)
[146] 2013, Catling, David C. *Astrobiology: A Very Short Introduction* Oxford University Press Kindle Edition. (L 319)
[147] 2011, Allen, Terence; Cowling, Graham, *The Cell: A Very Short Introduction* Oxford University Press Kindle Edition. (L 852)
[148] 2011, Allen, Terence; Cowling, Graham, *The Cell: A Very Short Introduction* Oxford University Press Kindle Edition. (L 383)
[149] 2011, Allen, Terence; Cowling, Graham, *The Cell: A Very Short Introduction* Oxford University Press. Kindle Edition. (L 515)
[150] 2011, Allen, Terence; Cowling, Graham, *The Cell: A Very Short Introduction* Oxford University Press. Kindle Edition. (L 286)
[151] 2013, Catling, David C. *Astrobiology: A Very Short Introduction* Oxford University Press Kindle Edition. (L 687)

[152] 2011, Allen, Terence; Cowling, Graham, *The Cell: A Very Short Introduction* Oxford University Press. Kindle Edition. (L 306)

[153] 2011, Allen, Terence; Cowling, Graham, *The Cell: A Very Short Introduction* Oxford University Press. Kindle Edition. (L 313)

[154] 2011, Allen, Terence; Cowling, Graham, *The Cell: A Very Short Introduction* Oxford University Press. Kindle Edition. (L 347)

[155] 2011, Allen, Terence; Cowling, Graham, *The Cell: A Very Short Introduction* Oxford University Press. Kindle Edition. (L 705)

[156] 2002, Polkinghorne, John, *Quantum Theory: A Very Short Introduction* Oxford University Press. Kindle Edition. (L 832)

[157] 2009, Blundell, Stephen J. *Superconductivity: A Very Short Introduction* Oxford University Press. Kindle Edition. (L 1078)

[158] 2009, Blundell, Stephen J. *Superconductivity: A Very Short Introduction* Oxford University Press. Kindle Edition. (L 1151)

[159] 2009, Blundell, Stephen J. *Superconductivity: A Very Short Introduction* Oxford University Press. Kindle Edition. (L 2112)

[160] 2012, King, Andrew, *Stars: A Very Short Introduction* Oxford University Press. Kindle Edition. (L 558) (p. 29)

The Auroral Oval

The word aurora can immediately bring to mind beautiful images of night skies filled with dancing streams of color. The aurora borealis (northern lights) and aurora australis (southern lights) have been widely photographed and their ability to create streaming images ranks as one of the wonders of the world. Beyond the colorful beauty of the downward and upwards cascading streams there is a vast amount of energy in motion.

This energy is caused by electrons flowing through the auroral oval into a volatile ionosphere. The auroral oval empties electrons into Earth's atmosphere interacting with all the atmospheric layers. Strong electron flow causes large auroral arcs where like honey poured on the top of Earth's sphere the auroral curtain travels down the sides at times almost reaching the equator.

Martin Redfern, Science Producer at the BBC says, "It is as if our world were an onion; a series of concentric spheres, from magnetosphere and atmosphere, through biosphere and hydrosphere, to the layers of the solid earth. Not all are spherical and some are much less substantial than others, but each manages to persist in a delicate equilibrium. Each component of such a system is seen not as something fixed and unchanging but more like a fountain; maintaining its overall structure perhaps, but constantly changing as material and energy pass through it." [161]

Dr. Syun-Ichi Akasofu, Professor of Physics Emeritus and Founding Director of the International Arctic Research Center of the University of Alaska Fairbanks says in his excellent and informative book Exploring the Secrets of the Auroral Oval,

Second Edition, "The magnetosphere is the region above the ionosphere in which the magnetic field of the Earth has the dominant control over the motions of gas and fast charged particles ... The Earth's electromagnetic environment is continuously monitored by recording changes of the Earth's magnetic field ... indicating that those changes occurred on a global scale ... geomagnetic storms are the manifestations of significant changes in the charge of the Earth's magnetic field." [162]

Professor Syun-Ichi Akasofu continues, "The ionosphere must be playing an active role in the dipolarization and perhaps even the over-dipolarization. Thus, the ionosphere actively participates in substorm processes, rather than passively responding to magnetospheric processes. [163] There can be more magnetic flux in the magnetotail during substorms than during a quiet period. [164] The magnetotail is simply the "tail" of the ionosphere." [165]

At the exosphere, in the upper region of the ionosphere approximately 600 kilometers above Earth, [166] electrons leave cold interplanetary space to enter Earth's warmer magnetosphere. Electrons are guided to the auroral oval by magnetic field lines. At the rate of 1.5 billion volts an hour electrons spiral through the westward turning ring current that forms the mouth of the negatively charged 10.4 kilometer wide northern auroral oval that delineates Earth's polar cap. The westward turning ring current parallels the path of the Sun. The downward flowing electrons become part of a magnetic sheet current that travels the stretched field-aligned magnetic currents.

The field-aligned currents flow between the magnetosphere and the ionosphere as a result of the magnetosphere-ionosphere coupling. [167] The upward sheet current which is carried by downward flowing electrons, is associated with an auroral arc. [168] The upward field-aligned current in a sheet form, carried by the downward streaming electrons, causes the curtain form of an auroral arc. [169]

Professor Stephen J. Blundell says, "The circulation of an electric field causes changes in a magnetic field. An electrical current causes the circulation of a magnetic field. The circulation of magnetic field causes a current or flow of charge and causes changes in an electric field. A changing magnetic field produces an electric field and a changing electric field produces a magnetic field." [170]

A geomagnetic storm occurs when intense substorms occur frequently. [171] Substorms are essential elements of a geomagnetic storm. [172] A geomagnetic storm field is produced by various electric current systems that develop around the Earth when solar disturbances reach the Earth. At high latitudes strong magnetic substorms occur during a geomagnetic storm where intense impulsive disturbances occur. [173]

The sudden storm commencement of a geomagnetic storm is caused by the impact of the shock wave on the magnetosphere. The shock wave is generated by a solar plasma/magnetic cloud advancing in the solar wind after being ejected during solar activities. The main phase is caused by the formation of a belt of energetic particles that surround Earth. This belt is called the ring current belt. [174] The shock wave compresses the magnetosphere. When the magnetic field in the plasma cloud has an intense southward component, it increases the dynamo power, causing a frequent occurrence of magnetospheric substorms and thus subsequently generating the ring current belt and the magnetospheric storm. [175]

Professor Syun-Ichi Akasofu says, "Substorms are the cause of the ring current belt, injecting high-energy protons from the magnetotail into that belt. Carl McIlwain and his colleagues (1974) showed that both protons and electrons are injected into the ring current belt and drift around the Earth...Oxygen ions of ionospheric origin become the dominant ions in Earth's ring current belt during an intense geomagnetic storm. Since the oxygen ions in the solar wind are highly ionized. A recent observation shows that ions are ejected out from the ionosphere into the magnetotail at substorm onset. After reaching the magnetotail, these ions are injected into the ring current belt by a convective motion of plasma in the magnetotail." [175a]

The auroral zone is a circular belt in the geomagnetic coordinate system, centered around the geomagnetic pole. [176] Auroral activity and geomagnetic disturbances are only different manifestations of an enhanced dynamo power. The discharge current system in the magnetosphere and the dynamo process feeds the current. A typical geomagnetic disturbance field undergoes a specific sequence of changes. The geomagnetic storm is the magnetic manifestation of a magnetospheric substorm. The magnetosphere has a specific response to increased power by the solar wind-magnetosphere dynamo for a few hours called magnetospheric substorms their manifestations being the polar magnetic and auroral substorm. Substorms are responsible for feeding oxygen ions from the ionosphere into the ring current. [177]

Physicists, Professor Lou Frank and Professor James Van Allen, and Research Scientist, John Craven plotted the outer boundary of the outer radiation belt onto the Earth's surface. [178] The boundary they delineated coincided fairly well with the auroral oval. This result suggested that auroral electrons penetrate into the polar upper atmosphere by moving along the outer boundary of the outer radiation belt. [179]

Dr. Al Zmuda, a Geophysicist and his colleagues (1966) found on the basis of TRIAD satellite data that field-aligned currents flow in or out from a belt that is basically identical to the auroral oval. [180] This fact suggested that auroral electrons carry field-aligned currents. [181] Auroral arcs appeared where there was upward field-aligned current. [182] Auroral arcs are found in a very specific belt, the auroral oval. [183] The width of the auroral oval changes intermittently. [184] The diffuse aurora is caused by energetic electrons from the outer radiation belt. [185]

Physicists, Professor Takeshi Iijima and Dr. Tom Potemra (1976) completed Dr. Zmuda's work by showing the distribution of field-aligned currents at the ionospheric level. Solar protons of energies on the order of 1.5 MeV (megaelectronvolts) were found to penetrate uniformly over the polar region bounded by the aurora oval. Energetic solar electrons were also found in the area bounded by the auroral oval. These results indicated that the auroral oval delineates approximately the boundary of the polar cap. The field lines that originate at the polar cap are connected with the interplanetary magnetic field lines. [186]

Professor Syun-Ichi Akasofu says, "I think it is important to note that the magnetosphere should be considered a system that converts the kinetic component of the solar wind energy into electromagnetic energy, since geomagnetic and auroral phenomena are various manifestations of electromagnetic energy dissipation processes. The magnetosphere must thus be a dynamo for this conversion. It transforms the kinetic (input) energy of the solar wind into substorm energy and eventually into heat (output) energy in the ionosphere. The southward component of the interplanetary magnetic field facilitates this energy transfer process. [187] The magnetopause is where the solar wind-magnetosphere dynamo is located." [188]

In 1960 Dr. Gene Parker, an Astrophysicist theorized that the Sun blows out plasma continuously with a supersonic speed from the whole solar surface under a certain temperature profile in the corona he coined the term solar wind. It was found that the Earth is surrounded by an extensive atmosphere of ionized gasses. Based on the study of atmospherics (radio emissions generated by thunderstorm lightning) L.R.O. Storey (1953) found that atmospherics can propagate approximately along the geomagnetic field lines from one hemisphere to the other. The propagation requires an extensive ionized atmosphere to a distance of several Earth radii. This ionized atmosphere has been named the plasmasphere. The ionosphere feeds the ionized gases to the plasmasphere. [189]

In 1968 Dr. Sam Bame, a Nuclear Physicist and a team at Los Alamos discovered the most extensive region of plasma called the plasma sheet, in the tail region of the magnetosphere. The magnetosphere is not an empty cavity but it consists of several plasma domains. In the 1970s solar wind-like plasmas were found well inside the boundary of the magnetosphere, and the region occupied by such plasmas is called the plasma mantel. The plasma in the plasma mantel flows in the anti-solar direction with a speed appreciably less than the solar wind. The plasma in the mantel is of solar wind origin. Oxygen ions of ionospheric origin instead of solar wind protons become the dominant ions in the ring current belt during intense geomagnetic storms. [190]

Professor Jim Dungey in 1961 suggested that the magnetic field lines carried by the solar wind are connected with some of the geomagnetic field lines across the boundary of the magnetosphere. [191] In 1931 Dr. Sydney Chapman, a Mathematician and Vincenzo Ferraro his Graduate Student introduced the concept of confinement of the Earth's magnetic field in a cavity carved in the solar gas flow. [192]

The solar wind stretches the dipolar field lines all the way to the outer boundary of the heliosphere where the solar wind interacts with interstellar gas. As the Sun rotates the stretched field develops a spiral structure. As the solar wind and its magnetic field are continuously changing, the power of the solar wind-magnetosphere dynamo varies as a result. [193]

The termination shock is where the solar wind slows from supersonic to subsonic speed and large changes in plasma flow direction and magnetic field orientation occur. [193a] Electrons play a new important role in the shock dynamics and thermodynamics. The electrons react on the shock electric field in a very specific way, leading to suprathermal nonequilibrium distributions of the downstream electrons. [193b]

Kristian Birkeland a Physicist viewed the interaction between the solar gas and the Earth's magnetic field in terms of motions of solitary charged particles in a dipole field. He set up an elaborate discharge chamber to study the trajectories of electrons around what he called a terrella. [194]

Dr. Chapman and Vincent Ferraro considered the solar gas to be consisting of an equal number of positive and negative particles (plasma in present terminology) and attempted to understand the behavior of the plasma flow as it approached a dipole field. They inferred that the solar plasma flow forms a comet-like structure around the Earth, extending in the anti-solar direction and confining the Earth and its magnetic field in it. [195]

Dr. John Foster, Associate Director of MIT's Haystack Observatory says the observations from space validate measurements from the ground. The combination of space and ground-based data give a highly detailed picture of a natural defensive mechanism in Earth's magnetosphere. "This higher-density, cold plasma changes about every plasma physics process it comes in contact with ... It slows down reconnection, and it can contribute to the generation of waves that, in turn, accelerate particles in other parts of the magnetosphere. So it's a recirculation process, and really fascinating." Dr. Foster likens this plume phenomenon to a "river of particles," and says it is not unlike the Gulf Stream, a powerful ocean current that influences the temperature and other properties of surrounding waters. On an atmospheric scale, he says, plasma particles can behave in a similar way, redistributing throughout the atmosphere to form plumes that "flow through a huge circulation system, with a lot of different consequences ... What these types of studies are showing is just how dynamic this entire system is." [196]

Geomagnetic disturbances can be defined as the magnetic manifestation of an increased level of the solar wind-magnetosphere interaction, resulting in an increased electric power output and currents from the solar wind-magnetosphere dynamo process. Magnetic fields generated by the resulting increased electric currents are defined as geomagnetic disturbance fields. [197]

Polar magnetic disturbances, ionospheric disturbances, X-ray bursts, VLF emissions, geomagnetic micropulsations, and other phenomena occur in harmony with auroral substorms. [198] A magnetospheric storm consists of a number of magnetospheric substorms. [199]

Magnetospheric Physicists began to pay attention to the concept of an open magnetosphere when Dr. A.L. Vampola (1971) detected solar electrons of about 400 KeV (kiloelectronvolts) uniformly over the entire polar region. [200] The open region is defined as the highest latitude region that is free from auroral electrons except for the polar rain that consists of the high-energy tail of electrons in the solar wind. [201] The open region expands during substorm activity. There is a poleward shift of the precipitation boundary and a large expansion of the hard electron precipitation region indicating a large poleward expanding and bulge. [202] The open magnetic flux is at least four times greater during substorm activity than during a quiet period. [203]

Dr. Chapman coined the terms the eastward electrojet for the current in the evening sector, and the westward electrojet for the current in the morning sector. [204] The westward electrojet extends into the evening sector along the auroral oval with the westward-traveling surge. [205]

Magnetospheric disturbances are various manifestations of the power generated by the solar wind-magnetosphere dynamo. The aurora can then be understood as the only visible manifestation of electrical discharge processes that are powered by the dynamo. Its output power is usually one million megawatts or more. The discharge takes place in the oval shaped belt called the auroral oval located in the polar upper atmosphere. [206]

Professor Syun-Ichi Akasofu says, "Without the concept of a medium (which now is known as solar plasma flow) that carries the effects of solar disturbances out into interplanetary space, it is not possible for the Sun to cause the magnetic changes recorded on the Earth." [207]

The auroral dynamo requires both the solar wind and a planetary magnetic field to be available. [208] A dynamo process in the solar photosphere generates the source of the energy for solar activities, since solar activities are basically electromagnetic phenomena. [209]

At 350 kilometers above the surface of the Earth electrons meet at the middle of the ionosphere, [210] an electromagnetic meeting place for highly energetic free electrons of varying strengths that split molecules apart and interact with neutral atoms and ions by changing their electric charge. This process is evidenced as electromagnetic substorms within the magnetic field of the auroral oval and ionosphere. Electron plasma separates releasing protons and freeing electrons.

The lower region of the magnetosphere is known as the ionosphere, a series of concentric shells of electrons and ions existing in regions of thin atmosphere. The lower layer is about 50 kilometers above the surface of the Earth and contains ionized molecules such as nitrogen and nitrogen monoxide. The upper layers extend up to about 500 kilometers and contain ionized atoms. Ultraviolet radiation from the Sun is responsible for ionizing the various layers and so the density of ions depends on whether it is day or night. [211]

The electrons in one atomic form or another continue downward passing out of the ionosphere's lower region called the thermosphere an area heated by ultraviolet electron energy. At 100 kilometers the temperature at the bottom of the thermosphere is -73 Celsius while at 300 kilometers the ionosphere temperature increases to 227 Celsius. [212] Alexander Oparin, Professor of Biology and Dr. John Haldane proposed that gases in Earth's early atmosphere were converted by ultraviolet sunlight or lightning into organic molecules. [213]

At 90 kilometers from the surface of Earth the electrons reach the mesosphere [214] an area where the cooler temperature causes the electrons to turn their energy inward insulating the newly acquired atomic nucleus. As the electrons cool they add to the electromagnetic field of Earth.

The electrons interact with Earth's high circulating noctilucent clouds composed of crystals of water ice. Teleconnections in Earth's atmosphere stretch all the way from the North Pole to the South Pole, linking weather and climate. Professor Cora Randall, Chair of the Dept. of Atmospheric and Oceanic Sciences at the University of Colorado says, "Stratospheric winds over the Arctic control circulation in the mesosphere ...When northern stratospheric winds slow down, a ripple effect around the globe causes the southern mesosphere to become warmer and drier, leading to fewer noctilucent clouds. When northern winds pick up again, the southern mesosphere becomes colder and wetter, and the noctilucent clouds return." [215]

At 50 kilometers above the surface of the Earth the electrons enter the stratosphere passing through the ozone layer [216] to circle the planet interacting with Earth's westward moving atmosphere.

At 10 kilometers above the surface the electrons reach the troposphere [217] where they becomes fully engaged with Earth's climatic weather patterns before touching land or water.

At 0 kilometers the electrons have reached ground level where combinations of orbiting electrons find partners and bond as pairs to produce molecular oxygen using the abundant oxygen atoms caused by photosynthesis where inorganic carbon in the form of carbon dioxide, affects the rate of the photosynthetic electron transport and thus the rate of photosynthetic oxygen production.

It is well known that inorganic carbon in the form of carbon dioxide is reduced in a light driven process known as photosynthesis to organic compounds. Less well known is that inorganic carbon also affects the rate of the photosynthetic electron transport and thus the rate of photosynthetic oxygen production. This result was first published by the Nobel Prize winner Otto Warburg in the late 1950s. [218]

Nearly all atmospheric oxygen is biological. The major source of oxygen is oxygenic photosynthesis, in which green plants, algae, and cyanobacteria use sunlight to split water into hydrogen and oxygen. These organisms combine the hydrogen with carbon dioxide to make organic matter, and they release oxygen. [219]

Oxygen's urge to engage in chemical reactions is excelled by only a very few other elements. [220] Oxygen is an extremely abundant element. It is the third most abundant element in the universe and the most abundant (47% of the total) in the Earth's crust. [221]

One of the most prominent emissions from the aurora is the greenish-white light from oxygen atoms. Most of the processes leading to the production of oxygen atoms are directly or indirectly related to molecular oxygen produced near ground level. Thus, the oxygen emission, the so-called green line of the terrestrial aurora, arises mostly because plants release abundant free oxygen into the atmosphere by photosynthesis. [222]

Professor Syun-Ichi Akasofu says, "It is expected that a number of stars are accompanied by several planets, and it may not be too long before the aurora on such planets can be discovered. [223] One possible way to detect plant life on such planets is to examine their auroral emissions. If strong oxygen emissions can be detected among other emissions in the planetary auroral oval, the possibility of the presence of plant life is high. Further, if plant life exists, animal life, whether lower or higher, can also exist there." [224]

Sources for The Auroral Oval

[161] 2003, Redfern, Martin, *The Earth: A Very Short Introduction* Oxford University Press. Kindle Edition. (L 188)
[162] 2007, Akasofu, Syun-Ichi, *Exploring the Secrets of the Aurora* Kindle. (L 158)
[163] 2007, Akasofu, Syun-Ichi, *Exploring the Secrets of the Aurora* Kindle. (L 1181)
[164] 2007, Akasofu, Syun-Ichi, *Exploring the Secrets of the Aurora* Kindle. (L 1216)
[165] 2007, Akasofu, Syun-Ichi, *Exploring the Secrets of the Aurora* Kindle. (L 1181)
[166] 2015, University of Waterloo;
http://www.science.uwaterloo.ca/~cchieh/cact/applychem/atmosphere.html

[167] 2007, Akasofu, Syun-Ichi, *Exploring the Secrets of the Aurora* Kindle. (L 1005)

[168] 2007, Akasofu, Syun-Ichi, *Exploring the Secrets of the Aurora* Kindle. (L 1121)

[169] 2007, Akasofu, Syun-Ichi, *Exploring the Secrets of the Aurora* Kindle. (L 1123)

[170] 2012, Blundell, Stephen J. *Magnetism: A Very Short Introduction* (L 775) Oxford University Press. Kindle Edition.

[171] 2007, Akasofu, Syun-Ichi, *Exploring the Secrets of the Aurora* Kindle. (L 589)

[172] 2007, Akasofu, Syun-Ichi, *Exploring the Secrets of the Aurora* Kindle. (L 589)

[173] 2007, Akasofu, Syun-Ichi, Exploring the Secrets of the Aurora Kindle. (L 167)

[174] 2007, Akasofu, Syun-Ichi, *Exploring the Secrets of the Aurora* Kindle. (L 167)

[175] 2007, Akasofu, Syun-Ichi, *Exploring the Secrets of the Aurora* Kindle. (L 237)

[175a] 2007, Akasofu, Syun-Ichi, *Exploring the Secrets of the Aurora* Kindle. (L 592)

[176] 2007, Akasofu, Syun-Ichi, *Exploring the Secrets of the Aurora* Kindle. (L 1027)

[177] 2007, Akasofu, Syun-Ichi, *Exploring the Secrets of the Aurora* Kindle. (L 218)

[178] 2007, Akasofu, Syun-Ichi, *Exploring the Secrets of the Aurora* Kindle. (L 723)

[179] 2007, Akasofu, Syun-Ichi, *Exploring the Secrets of the Aurora* Kindle. (L 724)

[180] 2007, Akasofu, Syun-Ichi, *Exploring the Secrets of the Aurora* Kindle. (L 728)

[181] 2007, Akasofu, Syun-Ichi, *Exploring the Secrets of the Aurora* Kindle. (L 730)

[182] 2007, Akasofu, Syun-Ichi, *Exploring the Secrets of the Aurora* Kindle. (L 732)

[183] 2007, Akasofu, Syun-Ichi, *Exploring the Secrets of the Aurora* Kindle. (L 701)

[184] 2007, Akasofu, Syun-Ichi, *Exploring the Secrets of the Aurora* Kindle. (L 713)

[185] 2007, Akasofu, Syun-Ichi, *Exploring the Secrets of the Aurora* Kindle. (L 1234)

[186] 2007, Akasofu, Syun-Ichi, *Exploring the Secrets of the Aurora* Kindle. (L 734)

[187] 2007, Akasofu, Syun-Ichi, *Exploring the Secrets of the Aurora* Kindle. (L 585)

[188] 2007, Akasofu, Syun-Ichi, *Exploring the Secrets of the Aurora* Kindle. (L 567)

[189] 2007, Akasofu, Syun-Ichi, *Exploring the Secrets of the Aurora* Kindle. (L 182)

[190] 2007, Akasofu, Syun-Ichi, *Exploring the Secrets of the Aurora* Kindle. (L 193)

[191] 2007, Akasofu, Syun-Ichi, *Exploring the Secrets of the Aurora* Kindle. (L 204)

[192] 2007, Akasofu, Syun-Ichi, *Exploring the Secrets of the Aurora* Kindle. (L 154)

[193] 2007, Akasofu, Syun-Ichi, *Exploring the Secrets of the Aurora* Kindle. (L 228)

[193a] 2015, http://voyager.jpl.nasa.gov/mission/interstellar.html

[193b] 2015, http://arxiv.org/abs/1505.02676

[194] 2007, Akasofu, Syun-Ichi, *Exploring the Secrets of the Aurora* Kindle. (L 151)

[195] 2007, Akasofu, Syun-Ichi, *Exploring the Secrets of the Aurora* Kindle. (L 155)

[196] 2014, Massachusetts Institute of Technology via Sciencedaily.com http://www.sciencedaily.com/releases/2014/03/140306142757.htm?utm_source=feedburner&utm_medium=email&utm_campaign=Feed%3A+sciencedaily%2Ftop_news%2Ftop_science+%28ScienceDaily%3A+Top+Science+News%29

[197] 2007, Akasofu, Syun-Ichi, *Exploring the Secrets of the Aurora* Kindle. (L 946)

[198] 2007, Akasofu, Syun-Ichi, *Exploring the Secrets of the Aurora* Kindle. (L 835)

[199] 2007, Akasofu, Syun-Ichi, *Exploring the Secrets of the Aurora* Kindle. (L 855)

[200] 2007, Akasofu, Syun-Ichi, *Exploring the Secrets of the Aurora* Kindle. (L 623)

[201] 2007, Akasofu, Syun-Ichi, *Exploring the Secrets of the Aurora* Kindle. (L 1194)

[202] 2007, Akasofu, Syun-Ichi, *Exploring the Secrets of the Aurora* Kindle. (L 1204)

[203] 2007, Akasofu, Syun-Ichi, *Exploring the Secrets of the Aurora* Kindle. (L 1209)

[204] 2007, Akasofu, Syun-Ichi, *Exploring the Secrets of the Aurora* Kindle. (L 969)

[205] 2007, Akasofu, Syun-Ichi, *Exploring the Secrets of the Aurora* Kindle. (L 1043)

[206] 2007, Akasofu, Syun-Ichi, *Exploring the Secrets of the Aurora* Kindle. (L 211)

[207] 2007, Akasofu, Syun-Ichi, *Exploring the Secrets of the Aurora* Kindle. (L 136)

[208] 2007, Akasofu, Syun-Ichi, *Exploring the Secrets of the Aurora* Kindle. (L 930)

[209] 2007, Akasofu, Syun-Ichi, *Exploring the Secrets of the Aurora* Kindle. (L 245)

[210] 2015, University of Waterloo;
http://www.science.uwaterloo.ca/~cchieh/cact/applychem/atmosphere.html

[211] 2012, Blundell, Stephen J. *Magnetism: A Very Short Introduction*
Oxford University Press. (L 1609) (p. 111) Kindle Edition.

[212] 2015, University of Waterloo;
http://www.science.uwaterloo.ca/~cchieh/cact/applychem/atmosphere.html

[213] 2007, Catling, David C. *Astrobiology: A Very Short Introduction*
Oxford University Press, Kindle Edition. (L 649)

[214] 2015, University of Waterloo;
http://www.science.uwaterloo.ca/~cchieh/cact/applychem/atmosphere.html

[215] 2014, NASA; http://science.nasa.gov/science-news/science-at-nasa/2014/16apr_teleconnections/

[216] 2015, University of Waterloo;
http://www.science.uwaterloo.ca/~cchieh/cact/applychem/atmosphere.html

[217] 2015, University of Waterloo;
http://www.science.uwaterloo.ca/~cchieh/cact/applychem/atmosphere.html
[218] 2014, Umeå universitet via SpaceDaily.com
http://www.sciencedaily.com/releases/2014/04/140413154053.htm?utm_source=feedburner&utm_medium=email&utm_campaign=Feed%3A+sciencedaily%2Ftop_news%2Ftop_science+%28ScienceDaily%3A+Top+Science+News%29
[219] 2007, Catling, David C. *Astrobiology: A Very Short Introduction* Oxford University Press Kindle Edition. (L 877)
[220] 2004, Ball, Philip, (L 694) *The Elements: A Very Short Introduction* Oxford University Press. Kindle Edition.
[221] 2004, Ball, Philip, (L 705) *The Elements: A Very Short Introduction* Oxford University Press. Kindle Edition.
[222] 2007, Akasofu, Syun-Ichi, Exploring the Secrets of the Aurora Kindle. (L 899)
[223] 2007, Akasofu, Syun-Ichi, Exploring the Secrets of the Aurora Kindle. (L 906)
[224] 2007, Akasofu, Syun-Ichi, Exploring the Secrets of the Aurora Kindle. (L 907)

Planets and Interplanetary Space

The planets in our solar system from the Sun out to Pluto although different in their size, temperature, rotation and orbital location all have electromagnetic ionospheric spheres. These spheres in some cases are very strong and in others are seemingly weak. But whatever their strength these spheres facilitate the flow and interaction of electrons that enter and leave a planet.

The point here is that electrons move cargo from planet to planet. The ionospheric actions of each planet determine how the cargo is accepted and released.

Today solar coronal mass ejections shoot plasma out into interplanetary space. The electrons in the plasma flow through the solar magnetic field and into the interplanetary magnetic field that links via magnetic field lines to the auroral ovals and ionospheres of planets.

The solar magnetic field extends well beyond the orbit of Pluto. The immense magnetic bubble that contains our solar system is called the heliosphere. The heliopause is the name for the blurred boundary between the heliosphere and the interstellar gas outside the solar system. [225] The meeting of the heliosphere and interstellar space occurs at the current sheet a sprawling surface within the heliosphere where the polarity of the Sun's magnetic field changes from plus to minus. [226]

Russell Stannard, OBE, Professor Emeritus of Physics says, "Space is no longer to be regarded as the passive stage on which the actors – material objects and light – perform their drama. Space itself becomes a performer." [227]

On a continuous basis all the planets in our solar system are releasing electrons out into interplanetary space. Electrons are continuously arriving at planet Earth. David C. Catling, Professor Earth and Space Sciences at University of Washington says, "Life is constructed from a limited toolkit, the periodic table of chemical elements, which is the same throughout the universe ... Because life has to get started and propagate, it's probable that the main atoms of life are abundant ones. Carbon is fourth in cosmic abundance after hydrogen, helium, and oxygen. Indeed, astronomers have found that many non-biological organic molecules already exist in space." [228] Interplanetary space appears to contain equal numbers of positively and negatively charged particles. [229]

Some electrons are blown by the steady force of the solar wind that is in fact an electron wind carrying mostly neutrons surrounded by mixed energy level electrons that have been set in motion by the high temperatures at the surface photosphere that radiates the light of the Sun. The solar wind plasma is the electron manifesting in one of its many forms. From the interior of the Sun very high energy electrons are ejected through the photosphere into the solar atmosphere where they burst into space as dense electron plasma clouds full of protons. [230]

The solar wind near our Sun's surface contains alternating streams of high and low speed. These streams rotate along with the Sun. The high-speed streams originate in coronal holes and extend toward the solar poles while the low-speed streams come from near the Sun's equator. With increasing distance from the Sun, the high-speed streams overtake the slower plasma, producing co-rotating interaction regions on their leading edges. [231]

The loops of the solar magnetic field that arch into the Sun's corona and are anchored in the active regions of low latitudes will be stretched out, near the ecliptic, into interplanetary space by the out-flowing solar wind. While these loops are pulled out radially, the Sun rotates about its axis, leading to the winding-up of the stretched-out field lines. As a result the interplanetary field assumes a spiral configuration. [232]

The continuous high temperature of the Sun causes outward radiation that accelerates into cold interplanetary space (-454 Fahrenheit). [233] The cold magnetic sea in interplanetary space becomes an ocean of motion so vast as to encompass our entire solar system. Through the commonality of electrons, infinite potentials are delineated to every location within and upon the magnetic sea, where any moment of time can be an experienced reality. Dr. Alan Hall says, "The Dirac Sea is an invisible field in space that is filled with virtual electrons which can become real by the addition of energy." [234]

In 1995, researchers discovered that if you took a few million rubidium atoms and cooled them near absolute zero, they would merge into a single wave of matter. [235] Peter Atkins, Professor of Chemistry says, "Interesting things happen to matter when it is cooled to very low temperatures. For instance, the original version of superconductivity, the ability of certain substances to conduct electricity with zero resistance, was discovered when it became possible to cool matter to the temperature about 4 Kelvin." (-452 Fahrenheit) [236]

The concept of negative temperature (thermodynamics below zero) applies only to systems that possess two energy levels. In a region of negative temperature, energy is conserved. The second law of thermodynamics implies that there will be a spontaneous transfer of heat from a system of negative temperature in contact with one of positive temperature and that the process will continue until the temperatures of the two systems are equal. The heat flows from the system with the lower (negative) temperature to the system with the higher (positive) temperature. [237] The available energy of a pure substance decreases as the temperature is raised. [238]

Professor Peter Atkins says, "Although there may be a change in internal energy of a certain value, the system in effect has to pay a tax to the surroundings in the sense that some of that change in internal energy must be used to drive back the atmosphere in order to make room for the products. [239] An actual example of a system that has only two energy levels is an electron spin in a magnetic field. [240] As the temperature is raised, electrons migrate into the upper state." [241]

When the electron travels through cold interplanetary space it turns inward to create energy ready for release at a later time when the electron arrives at a higher temperature location. As the electron internally creates energy it also externally creates a bed of electromagnetic energy that when accompanied with other electrons manifests as waves that appear in such abundance as to form a sea that is known as the interplanetary magnetic field.

Dr. Percy Seymour says, "Associated with the interplanetary magnetic field is a sheet of electric current. The component of the interplanetary field, that is perpendicular to the ecliptic, changes sign when the Earth passes through the current sheet as it orbits the Sun. When this happens, charged particles can more easily enter the Earth's ionosphere, because charged particles preferentially travel along field lines." [242]

Mathematical Physicist James C. Maxwell realized that changes in an electric field will produce changes in magnetic field, and vice versa, and a self-sustaining wave of varying electric and magnetic fields will propagate off into space. Maxwell had predicted the existence of an electromagnetic wave with a speed equal to the speed of light. [243] Magnetism is another link between us and the rest of the universe. Magnetic fields influence the particle and magnetic environment of our Earth. [244]

Electrons interacting with the interplanetary magnetic field approach Earth where at the exosphere the electrons leave cold interplanetary space to enter our planet's warmer magnetosphere that starts at the upper region of the ionosphere at a circumpolar belt called the auroral oval. [245] Auroral substorms are associated with a direction change of the interplanetary magnetic field caused by a positive value changing to a negative value. [246]

The motions of the planets in their orbits coupled with the magnetospheric inflows and outflows at the magnetic poles facilitated by an auroral oval creates currents of varying temperatures within the interplanetary magnetic field where waves of electrons move as fast as the speed of light through every part of our solar system.

Professor Russell Stannard says, "As an object increases its speed so it also increases its energy; it acquires kinetic energy – energy of motion. Energy is assumed to possess mass ... So what we are saying is that the total energy of the object is the sum of the energy locked up in the rest mass of the object, plus the kinetic energy." [247]

In 1755, Immanuel Kant suggested that the solar system coalesced out of a diffuse cloud in space. [248] David A. Rothery, Professor of Planetary Geosciences says, "Exploration of our solar system has reached a stage that allows us to appreciate other planets and their large satellites as worlds, endowed with geographies, geologies, and meteorologies as complex and fascinating as those of our own planet, Earth. [249] ...After its birth, each terrestrial planet must have developed an atmosphere when internal gases leaked out via the magma ocean." [250]

The stronger gravity of the more massive terrestrial planets enables them to hang on to gas more effectively, although the density and chemical composition have evolved out of all recognition as a result of numerous processes. An important ongoing process is that short-wavelength solar ultraviolet light can split molecules of water vapour into hydrogen and oxygen. Hydrogen is very light, and can escape to space. [251]

Deeper layers of an atmosphere, where the short-wavelength ultraviolet does not penetrate, are immune from photochemistry. Here the air is warmed mostly by contact with the ground (which is warmed directly by the Sun). In the lowest layer, called the troposphere, atmospheric temperature decreases upwards. [252]

The use of space probes to explore the environments of the planets have shown that other planets that have magnetic fields also have magnetospheres with similarities to Earth's. [253] The magnetic field of a planet is cocooned inside a zone called the magnetosphere. Electron controlled charged particles can be channelled along magnetic field lines towards the top of the atmosphere causing auroral glows that are well known on Earth and observed also on Jupiter and Saturn. [254]

Earth's magnetosphere has a tear drop shape, extending about 10 Earth radii in the direction towards the Sun, and possibly a couple of hundred Earth radii in the direction away from the Sun. The magnetosphere forms a protective layer around the Earth, cocooning the planet and providing some protection from the harsh environment of the solar wind. [255]

Each of the giant planets has a strong magnetic field. The magnetic dipole moment' of Neptune, which is the conventional measure of a planetary magnetic field, is 25 times greater than the Earth's. Uranus' is 38 times, Saturn's is 582 times, and Jupiter's is 1,949 times greater. [256]

Professor David A. Rothery says, "The internal structure of the giant planets may still be evolving because, with the possible exception of Uranus, they all radiate more heat to space than they receive from the Sun. Jupiter is so massive that it could still be leaking out a significant amount of primordial heat trapped within since its formation, but for Saturn and Neptune this heat excess shows that heat must be generated within the planet." [257]

Professor David A. Rothery continues, "Probably, the troposphere of each giant planet merges seamlessly into a fluid interior at temperatures and pressures so high that there is no distinction between gas and liquid. Certainly, there is no solid surface that a human could ever stand upon." [258]

Professor David C. Catling says, "I will argue that at least nine other bodies in our solar system might be habitable today, if we keep an open mind. Solar system astrobiology is far from settled." [259]

Inner Planets

The Earth's atmosphere differs from nearby planets in the complexity of its layering. The Earth is unique among terrestrial planets in having a layer extending from about 10 to 50 kilometres altitude, between its troposphere and mesosphere, where temperature increases with altitude. This is the stratosphere, which is warmed by the absorption of ultraviolet photons by ozone molecules. [260]

The surface of a planet is warmer than it would be in the absence of an atmosphere because it receives energy from a heated atmosphere in addition to visible sunlight. [261] The temperature of the atmosphere depends on two factors: (1) the heat it picks up from contact with the ground; (2) how much of the outgoing infrared radiation it can absorb. [262] Earth's stratospheric ozone layer of concentrated ozone between 20 and 30 kilometres in height shields the Earth's surface from harmful ultraviolet rays. [263]

Professor David A. Rotherby says, "Atmospheric molecules split by ultraviolet light can combine with others, by series of reactions described as photochemistry. This occurs especially in a planet's thermosphere which begins about 100 kilometres above the surface. The thermosphere is heated by the solar ultraviolet energy used in either splitting molecules apart or stripping them of some of their electrons. The latter process is called ionization, and ions (mainly of oxygen for the Earth and carbon dioxide for Venus and Mars) can be sufficiently common in the outer reaches of a thermosphere to make an electrically conducting layer referred to as the 'ionosphere'. When a solar storm brings plasma from the Sun to the Earth, this distorts the magnetic field and causes unusual currents to flow in the ionosphere." [264]

The atmosphere warms the Earth to an extent depending on atmospheric composition. Without an atmosphere the Earth's surface would be −18 Celsius. Instead, today's average global surface temperature is +15 Celsius. The size of The Earth's greenhouse effect which is the warming caused by the atmosphere is 33 Celsius the difference between -18 Celsius and +15 Celsius. [265] The tropospheric temperature of a planet is controlled by how effectively the lower atmosphere absorbs electromagnetic radiation. This is because visible sunlight warms the ground, and the warmed ground emits electromagnetic radiation. [266]

The Earth's hydrologic cycle is the interplay between interior, surface, and atmosphere, and the cycling of components between them and is extremely important. It is not a single cycle, but an array of interconnected loops. In essence, water in the oceans evaporates to form clouds, and later precipitates out as rain or snow that eventually finds its way back into the oceans (via rivers or seasonal polar caps). [267] Professor Philip Ball says, "The chemical constancy of our planet's environment is due not to inactivity but to perpetual change." [268]

At present, the Earth is hit annually by about 10,000 meteorites greater than 1 kilogram, but most of those are too small to survive passage through the atmosphere, where they are heated and worn away by friction. The yearly supply of 1,000 kilogram meteorites is only about 10. [269] A 1.9 kilogram piece of rock that had been blasted off the surface of Mars by an asteroid impact eventually landed in Antarctica. [270]

On airless bodies like the Moon and Mercury polar caps were not expected. However, during the 1990s it was noted that radar signals are reflected with unusual strength from permanently shadowed regions inside craters near the poles of both bodies. [271] The Moon and Mercury have too little gravity to hold gas atmospheres but they each have an exosphere a very tenuous outermost atmospheric zone. [272]

Dr. Patrick Peplowski, a research scientist at the Johns Hopkins University Applied Physics Laboratory and NASA Messenger team member said, "When Nasa's Messenger spacecraft arrived at Mercury we found unexpectedly high abundances of the moderately volatile elements [potassium] and [sulfur]...Since then, we've also observed the high concentrations of [sodium] and [chlorine], which were also expected to be depleted on Mercury." [273]

Conditions that exist on Earth's Moon cause interaction. Professor Russell Stannard says, "[on Earth] ... we have to contend with air resistance, which tends to slow down some falling objects more than others. Whereas a hammer falls directly down, a feather released at the same time will float down more leisurely. But when the effects of air resistance are excluded – as was the case when the astronauts on the Apollo 15 mission performed this experiment on the Moon – the feather and a hammer arrive at the ground at the same instant." [274] This experiment is based on Galileo's famous experiment.

The Moon's crust averages about 70 kilometres in thickness (13% of the Moon's volume), ranging from 100 kilometres in some highland regions to 20 kilometres below some major impact basins. The Earth's oceanic crust is 6-11 kilometres thick, whereas continental crust varies from about 25 kilometres in thin, stretched regions to 90 kilometres below major mountain ranges. In total, crust occupies only about 1% of the Earth's total volume. [275]

Venus has a 2,000 kilometer diameter south polar vortex with clouds 60 kilometers above the surface that are drawn downwards by warmer temperatures. [276] Venus also has a dense, permanently cloudy atmosphere [277] that is a much denser atmosphere than Earth, with 92 times greater ground-level atmospheric pressure. Even sluggish winds can shift sand particles around. Venus has several fields of sand dunes. [278]

On Venus clouds form when the temperature and pressure make it favourable for some of the atmosphere to condense as liquid droplets or ice particles of usually water. Although water is only a small fraction of Venus' atmosphere, there is enough to form a continuous layer of cloud at the top of its troposphere, between about 45 and 65 kilometres above the surface. There, water vapour condenses as droplets and remains suspended as aerosol droplets too small to fall. Atmospheric sulfur dioxide dissolves in them turning them into sulfuric acid. Wherever the droplets are drawn down below about 45 kilometres by atmospheric circulation, the heat causes them to evaporate again. [279]

Professor David C. Catling says, "It is believed that Venus lost its oceans due to runaway greenhouse effects that caused the ocean to sweat into space." [280] There is a natural greenhouse effect in the atmospheres of Venus, Earth, and Mars. Thanks mostly to its enormous load of carbon dioxide the atmospheric greenhouse effect on Venus maintains its surface temperature an impressive 500 Celsius above what it would otherwise be. Water vapour and carbon dioxide warm the Earth by about 30 Celsius, and greenhouse warming of Mars, which has a tenuous carbon dioxide-rich atmosphere, is only about 6 Celsius. [281] On Venus this dense blanket of the greenhouse gas raises surface temperatures to around 750

Celsius while on Mars the thin sheet keeps things at a frigid – 50 Celsius. [282]

At Venus and Mars, temperature decreases rapidly with altitude in the troposphere, then decreases more slowly with altitude in the mesosphere, and then rises with altitude in the thermosphere because of the ultraviolet absorption. [283] Convection occurs when air near the base of the troposphere is heated and expands making it buoyant. It will then rise, to be replaced by colder air displaced from above. This is what drives the weather on the Earth, Venus, and Mars. The pattern of circulation is different in each case because it also depends on such factors as the planet's rate of rotation (slow for Venus), the rate of rotation of the atmosphere (much faster than that of the planet itself in the case of Venus' upper troposphere), and the day-night temperature difference (high for Mars, small for Venus). [284]

Professor Philip Ball says, "Venus and Mars are of a similar size to Earth, and they were formed from a roughly similar mixture of elements. But their skies now contain only tiny amounts of oxygen – less than 1% – and only small quantities of nitrogen. Their atmospheres are both about 95 per cent carbon dioxide, even though that of Mars is very tenuous while that of Venus is very thick." [285]

On Venus clouds are highly reflective, so the cloudier an atmosphere, the more solar energy is reflected directly back into space. However, a cloudy sky increases the ability of an atmosphere to trap heat from the sunlight that does reach the ground, so the effect of clouds on global temperature is complex. The unbroken clouds of Venus have not saved its surface from being thoroughly cooked by the greenhouse effect. [286]

The magnetic environment of Venus was studied by a series of Mariner and Veneras space probes and also by the Pioneer Venus Orbiter. The investigations showed that the strength of the dipolar field of Venus is less than 1/10,000 of Earth's field. Eddy currents induced by the conducting ionosphere of Venus by the solar wind prevent the wind from reaching its surface.

Dr. Percy Seymour says, "The use of space probes to explore the environments of the planets have shown that other planets that have magnetic fields also have magnetospheres with similarities to Earth's." [287]

Underscoring the vast differences between Earth and Venus, new research (September 2014) shows a glimpse of giant holes in the electrically charged layer of the Venusian atmosphere, called the ionosphere. The observations point to a more complicated magnetic environment than previously thought. Venus has an ionosphere, a layer of the atmosphere filled with charged particles. The Venusian ionosphere is bombarded on the Sun-side of the planet by the solar wind. Consequently, the ionosphere, like air flowing past a golf ball in flight, is shaped to be a thin boundary in front of the planet and to extend into a long comet-like tail behind. As the solar wind plows into the ionosphere, it piles up like a big plasma traffic jam, creating a thin magnetosphere around Venus. [288]

Professor David A. Rothery says, "Mars is sure to have hydrologic cycles similar to Earth that act more sporadically and over different timescales, and with different relative importances for each loop. There are probably even slower cycles involving carbon dioxide and sulfur dioxide on Venus, in which the atmosphere weathers the surface rocks, which eventually become buried by lava flows to depths at which the gases are liberated once more and escape back to the atmosphere through volcanic vents. Until we have explored and documented the complexities and timescales of these multi-looped and inter-related cycles, our understanding of what makes each planet 'tick' will remain naive." [289]

Wind-blown sand is a powerful agent of erosion on Mars where some exposed layers of rock have become curiously sculpted by abrasion. The low atmospheric density means that a wind capable of transporting sand grains has to be blowing much faster than on Earth. [290]

Earth's clouds above about 6 kilometres consist mostly of tiny ice particles, and below that altitude they are mostly droplets of water. On Mars, clouds are comparatively rare. In most of its troposphere, they are made of water-ice, and around 80 kilometres, near the troposphere/mesosphere boundary, clouds of carbon dioxide particles have been observed. [291]

The Cassini mission to Mars found lakes of ethane-tainted liquid methane near both poles. [292] Professor David A. Rothery says, "There is no doubt that ice exists at the surface today in Mars' polar caps in the form of permanent water-ice with a fringe of carbon dioxide frost that grows and contracts seasonally." [293]

Professor David C. Catling says, "Part of the early Martian atmosphere was probably blasted away to space by large comet and asteroid impacts in so-called impact erosion. The atmosphere was vulnerable because of Mars's low gravity. Along with more gradual escape of gases to space later, Mars was ultimately left with its thin atmosphere." [294]

The icy nucleus of comet Siding Spring, about 700 meters across, made its closest approach to Mars on Oct. 19, 2014 passing within 86,000 miles. Close enough that gas and dust in the outermost reaches of the comet's atmosphere interacted with the atmosphere of Mars. [295]

Professor David C. Catling says, "Even the simplest microbes native to Mars or to Europa's oceans would change the balance of probabilities that life exists elsewhere in the galaxy for they would demonstrate that life can originate twice within one solar system." [296] "Perhaps originate should really be take hold".

The Sun

Dr. Alan Hall says, "Inside the Sun ongoing fusion reaction melts matter into plasma liberating electrons." [297]

Professor Stephen J. Blundell says, "Particles in plasma on the Sun are charged and moving very fast and able to produce and interact with a magnetic field via magnetic field lines that can become trapped by regions of fluid and dragged along with the fluid as it flows around. These field lines can become twisted and tangled because of the complex motion of the fluid." [298]

Claudio Tuniz, Professor at the International Center for Theoretical Physics says, "At the temperatures required for fusion, the hydrogen mass becomes a plasma, a state of matter in which atomic electrons are separated, creating an electrically charged mixture of positive ions and electrons. Like gas, a plasma does not have a definite shape or volume, but being charged, it can be shaped by magnetic fields." [299]

The Sun contains ionized gas (the outer electrons are ripped off atoms leaving the remainder positively charged) where the individual particles have an electrical charge, and the mix of ions and electrons forms a plasma. [300]

Solar flares sometimes occur on the surface of the Sun. These are enormous explosions that heat gas to many millions of degrees, emitting X-rays and gamma rays as well as charged particles. In coronal mass ejections, bubbles of gas which are threaded with magnetic field lines are expelled from the Sun over a period of several hours. Sometimes these violent events can disturb the Earth's magnetosphere when the charged particles arrive at Earth and cause a geomagnetic storm. [301]

The Hale-Nicolson laws state: (1) All pairs of sunspots in the same hemisphere have opposite polarity with regard to the leading and trailing members of the pair. If the leading member is a south magnetic pole, then the trailing member will be a north magnetic pole, but the polarities will be opposite from those in the other hemisphere; (2) During the next solar cycle the polarities in the two hemispheres will be reversed. This second law implies that the solar cycle is roughly 22 years long, rather than 11. [302]

Professor Stephen J. Blundell says, "At the beginning of each sunspot cycle, it is found that the sunspots begin to appear in two bands, each at relatively high latitudes (high up in the northern and low down in the southern hemispheres), but towards the end of the cycle the two bands are found closer to the equator. Understanding the solar dynamo requires use of the principles of fluid dynamics, plasma physics, and the interaction between lines of magnetic fields twisting and turning in a rotating, bubbling cauldron of convecting turbulent fluid." [303]

When sunspots occur in pairs, the magnetic field emerges from below the Sun's surface at one spot, and enters the surface again at the other spot. During one 11 year sunspot maximum, the spots in the two hemispheres had different east-west polarities from each other, but during the next cycle the polarities in each hemisphere would be reversed. Later work showed that the Sun has fields of opposite polarity at its two poles, and that this polarity changes at the start of a new sunspot cycle. Magnetic fields are also found in prominences seen against the disc, and quite often these occur between the magnetic areas of opposite polarity associated with sunspot pairs. [304]

The solar wind is the continual, almost radial, out flowing of the solar corona. Magnetic lines of force diverge from coronal holes. The coronal temperature is so high that the gravitational field of the Sun cannot contain this highly ionized gas in a confined static atmosphere. Two components of the solar wind have been identified: (1) a slow-moving low density flux; (2) superimposed on this, are high-speed streams. This latter emission is greatly increased over coronal holes. [305]

Dr. Percy Seymour says, "...there is considerable evidence to show that the movements and positions of the planets do have an effect on the solar cycle and on violent events on the Sun." [306]

Professor Andrew King says, "A star like the Sun will eventually expand to about two hundred times its present size, just about reaching the Earth's orbit." [307]

Outer Planets

In 2002 Professor Syun-Ichi Akasofu said, "There is some indication that Saturn's magnetosphere has substorm activities." [308] In 2009, Cassini encountered Saturn's auroral region at a distance of 247 million kilometres from Saturn's cloud tops. High above the spectacular visible-light displays of Saturn's northern and southern lights, auroral emissions occur this far from the planet at radio wavelengths. The emissions are generated by fast moving electrons spiralling along Saturn's magnetic field lines, which are threaded through the auroral region. [309]

The Hubble Space Telescope depicted a clear image of the auroral oval on Jupiter and Saturn. The solar wind-planetary magnetic field interaction is essential in providing the power for auroral discharge, as is the case for the Earth. [310]

Cassini while crossing Saturn's magnetic field lines that connect to the aurora obtained ultraviolet images of the aurora which manifests itself as a complete oval around each pole of the planet. Variations in the aurora provide information on changes in the associated magnetosphere. [311]

Saturn, like all magnetized planets, emits radio waves into space from the polar regions. On Saturn the radio emissions pulse while the auroras beat in tandem with the radio waves. This confirms that the auroras and the radio emissions are physically associated implying that the pulsing of the radio emissions is being imparted by the processes driving Saturn's aurora. [312]

Earth also has radio auroral emissions and September 2010 results show that the process that generates radio aurora appears to be the same at both Earth and Saturn. Interestingly, there are two minor differences between the aurora at Earth and Saturn. At Earth, there is a cavity in the plasma above the auroral oval that rises for several thousand kilometres. New observations show that this is not seen at Saturn. Secondly, radio sources were crossed at much further distances from the planet. These discrepancies reflect intrinsic differences between the two magnetospheres, in terms of dimensions and planetary rotation speed. It is thought that the unusual conditions responsible for these intense electric currents might have been triggered by a solar wind compression squeezing Saturn's magnetic field and producing the observed auroras. [313]

Saturn despite its remoteness still feels the Sun's influence. The Sun constantly emits particles that reach all the planets of the solar system as the solar wind. When this electrically charged stream in interplanetary space gets close to a planet with a magnetic field, like Saturn or the Earth, the field traps the particles, bouncing them back and forth between its two poles where they interact with atoms in the upper layers of the atmosphere, creating aurora. There are subtle differences between the northern and southern aurora of Saturn. The northern auroral oval is slightly smaller and more intense than the southern one, implying that Saturn's magnetic field is not equally distributed across the planet; it is slightly uneven and stronger in the north than the south. As a result, the electrically charged particles in the north are accelerated to higher energies as they are fired toward the atmosphere than those in the south. [314]

Aurora on Saturn occurs in a process similar to Earth's northern and southern lights. Particles from the solar wind are channeled by Saturn's magnetic field toward the planet's poles, where they interact with electrically charged gas (plasma) in the upper atmosphere and emit light. At Saturn, however, auroral features can also be caused by electromagnetic waves generated when the planet's moons move through the plasma that fills Saturn's magnetosphere. [315]

Data from NASA's Cassini spacecraft have revealed that Enceladus, one of Saturn's small moons, is linked to Saturn by powerful electrical currents -- beams of electrons that flow back and forth between the planet and moon. Research has found that jets of gas and icy grains emanate from the south pole of Enceladus, which become electrically charged and form an ionosphere. The motion of Enceladus and its ionosphere through the magnetic bubble that surrounds Saturn acts like a dynamo, setting up the current system. [316]

Jupiter is linked to three of its moons by charged current systems set up by the satellites orbiting inside its magnetosphere, and that these current systems form glowing spots in the planet's upper atmosphere. The latest discovery at Enceladus shows that similar processes take place at the Saturnian system too looking like a universal process. Jupiter's moon Io is the most volcanic object in the solar system, and produces a bright spot in Jupiter's aurora. Now, we see the same thing at Saturn -- the variable and majestic water-rich Enceladus plumes, probably driven by cryovolcanism, cause electron beams which create a significant spot in Saturn's aurora too. [317]

While NASA's Hubble Space Telescope, orbiting around Earth, was able to observe the northern auroras in ultraviolet wavelengths, NASA's Cassini spacecraft, orbiting around Saturn, got complementary close-up views in infrared, visible-light and ultraviolet wavelengths. Cassini could also see northern and southern parts of Saturn that don't face Earth. A new video showing aurora images from Hubble and Cassini is available at NASA. [318]

Dr. Jonathan Nichols of the University of Leicester in England, who led the work on the Hubble images said, "In 2013, we were treated to a veritable smorgasbord of dancing auroras, from steadily shining rings to super-fast bursts of light shooting across the pole. Images from Cassini's ultraviolet imaging spectrometer (UVIS), obtained from an unusually close range of about six Saturn radii, provided a look at the changing patterns of faint emissions on scales of a few hundred miles and tied the changes in the auroras to the fluctuating wind of charged particles blowing off the Sun and flowing past Saturn. The UVIS images, also suggest that one way the bright auroral storms may be produced is by the formation of new connections between magnetic field lines. That process causes storms in the magnetosphere around Earth. The movie [video] also shows one persistent bright patch of the aurora rotating in lockstep with the orbital position of Saturn's moon Mimas. While previous UVIS images had shown an intermittent auroral bright spot magnetically linked to the moon Enceladus, the new movie [*video*] suggests another Saturn moon can influence the light show as well. Dr. Sarah Badman, a Cassini visual and infrared mapping spectrometer team associate at Lancaster University said, "Scientists have wondered why the high atmospheres of Saturn and other gas giants are heated far beyond what might normally be expected by their distance from the Sun...By looking at these long sequences of images taken by different instruments, we can discover where the aurora heats the atmosphere as the particles dive into it." [319] More information about Cassini is available at NASA.gov. [320]

Jupiter and Saturn have a yellowish cast while Uranus and Neptune appear bluey-green. This is because we see their cloud-tops through a depth of overlying methane gas, which preferentially absorbs the longer (red) wavelengths of light. [321]

The topmost layer of continuous clouds on Uranus and Neptune is of methane-ice particles. On Jupiter and Saturn ammonia-ice particles condense to form the topmost clouds. These top cloud layers are about 10 kilometres thick, below which the air probably becomes clear again. [322] Calculations suggest that in the case of Jupiter, there should be a second layer of clouds made of ammonium hydrosulfide about 30 kilometres below, and a third cloud layer, this time of water (ice at the top, liquid water droplets below) about 20 kilometres lower still. The same cloud layers are expected at Saturn, but spaced about three times further apart because of Saturn's lower gravity. Ammonia-bearing clouds are expected below the methane clouds of Uranus and Neptune. [323] Pressures are too low for metallic hydrogen in Uranus and Neptune, so their magnetic fields are harder to account for, but are probably caused by motion within electrically conducting ice of their outer cores. [324]

At the speed of light, it would take 8.3 minutes to travel from the Earth to the Sun and a little over four hours to then travel to the average orbital distance of Neptune. [325] When Neptune was revealed in detail during Voyager 2's 1989 fly-by, it resembled a blue version of Jupiter. There was even a giant storm system in the form of a dark spot just south of the equator that proved to be short lived, and vanished by 1994. Unlike at Jupiter and Saturn, the equatorial wind stream on Neptune blows west (opposite to the planet's rotation). [326]

All planets have temperature changes related to their seasons with some changes more extreme than others. When Pluto is closest to the Sun it is at its warmest and can reach temperatures of - 369 Fahrenheit. When Pluto is farthest from the Sun it is at its coolest, temperatures can fall to - 387 Fahrenheit. [327]

New Horizons is the fastest spacecraft the United States has ever launched. It took the Apollo astronauts around three days to get to the Moon. New Horizons flew past the Moon in just eight hours on its journey to the Oort cloud that surrounds our solar system. New Horizons was launched Jan. 19, 2006 and flew past Pluto July 15, 2015 – meaning it takes 9.5 years to get to Pluto, even at that fantastic speed. To get the probe there in an amount of time that's reasonable it has to go very fast, and that means it will go by Pluto very fast as it continues on to the Oort cloud. [328] The early images so far show little more than Pluto's shape (spherical) and color (reddish). Over the years, changes in those color patterns hint at a dynamic planet where something is happening. At its closest approach July 2015, New Horizons was 10,000 kilometers above the surface of Pluto. [329]

Sources for Planets and Interplanetary Space

[225] NASA; http://cosmicopia.gsfc.nasa.gov/heliosph.html
[226] 2014, American Geophysical Union via ScienceDaily.com; http://www.sciencedaily.com/releases/2014/07/140723114142.htm
[227] 2008, Stannard, Russell, Relativity: A Very Short Introduction Oxford University Press, Kindle Edition. (L 1140) (p. 75)
[228] 2007, Catling, David C. Astrobiology: A Very Short Introduction
Oxford University Press, Kindle Edition. (L 348 & 354)

[229] 2012, King, Andrew, *Stars: A Very Short Introduction* Oxford University Press, Kindle Edition. (L 371) (pp. 16-17)

[230] 2014, New Jersey Institute of Technology via ScienceDaily.com http://www.sciencedaily.com/releases/2014/06/140603182609.htm?utm_source=feedburner&utm_medium=email&utm_campaign=Feed%3A+sciencedaily%2Ftop_news%2Ftop_science+%28ScienceDaily%3A+Top+Science+News%29

[231] NASA; http://cosmicopia.gsfc.nasa.gov/heliosph.html

[232] 2008, Seymour, Percy *Dark Matters: Unifying Matter, Dark Matter, Dark Energy, and the Universal Grid* (L1003) Kindle Edition.

[233] 2014, NASA; http://science.nasa.gov/science-news/science-at-nasa/2014/30jan_coldspot/

[234] 2011, Hall, Alan *Electron* (L 339-341) Kindle Edition.

[235] 2014, NASA; http://science.nasa.gov/science-news/science-at-nasa/2014/30jan_coldspot/

[236] 2010, Atkins, Peter *The Laws of Thermodynamics: A Very Short Introduction* Oxford University Press Kindle Edition. (L1424)

[237] 2010, Atkins, Peter *The Laws of Thermodynamics: A Very Short Introduction* Oxford University Press Kindle Edition. (L1574)

[238] 2010, Atkins, Peter *The Laws of Thermodynamics: A Very Short Introduction* Oxford University Press Kindle Edition. (L1304)

[239] 2010, Atkins, Peter *The Laws of Thermodynamics: A Very Short Introduction* Oxford University Press Kindle Edition. (L1171)

[240] 2010, Atkins, Peter *The Laws of Thermodynamics: A Very Short Introduction* Oxford University Press Kindle Edition. (L1523)

[241] 2010, Atkins, Peter *The Laws of Thermodynamics: A Very Short Introduction* Oxford University Press Kindle Edition. (L1529)

[242] 2008, Seymour, Percy *Dark Matters: Unifying Matter, Dark Matter, Dark Energy, and the Universal Grid* (L1012) Kindle Edition.

[243] 2012, Blundell, Stephen J. *Magnetism: A Very Short Introduction* Oxford University Press. Kindle Edition. (L 775)

[244] 2008, Seymour, Percy *Dark Matters: Unifying Matter, Dark Matter, Dark Energy, and the Universal Grid* (L1230) Kindle Edition.

[245] 2007, Akasofu, Syun-Ichi, *Exploring the Secrets of the Aurora* Kindle. (L 656)

[246] 2007, Akasofu, Syun-Ichi, *Exploring the Secrets of the Aurora* Kindle. (L 562)

[247] 2008, Stannard, Russell *Relativity: A Very Short Introduction* Oxford University Press. Kindle Edition. (L649) & (L683)

[248] 2007, Catling, David C. *Astrobiology: A Very Short Introduction* Oxford University Press Kindle Edition. (L 508)

[249] 2010, Rothery, David A. *Planets: A Very Short Introduction* Oxford University Press. Kindle Edition. (L 310)

[250] 2010, Rothery, David A. *Planets: A Very Short Introduction* Oxford University Press. Kindle Edition. (L 1084)

[251] 2010, Rothery, David A. *Planets: A Very Short Introduction* Oxford University Press. Kindle Edition. (L 1091)

[252] 2010, Rothery, David A. *Planets: A Very Short Introduction* Oxford University Press. Kindle Edition. (L 1106)

[253] 2008, Seymour, Percy *Dark Matters: Unifying Matter, Dark Matter, Dark Energy, and the Universal Grid* (L 464) Kindle Edition.

[254] 2010, Rothery, David A. *Planets: A Very Short Introduction* Oxford University Press. Kindle Edition. (L 1348)

[255] 2012, Blundell, Stephen J. *Magnetism: A Very Short Introduction* Oxford University Press. Kindle Edition. (L 1603)

[256] 2010, Rothery, David A. *Planets: A Very Short Introduction* Oxford University Press. Kindle Edition. (L 1333)

[257] 2010, Rothery, David A. *Planets: A Very Short Introduction* Oxford University Press. Kindle Edition. (L 1266)

[258] 2010, Rothery, David A. *Planets: A Very Short Introduction* Oxford University Press. Kindle Edition. (L 1237)

[259] 2007, Catling, David C. *Astrobiology: A Very Short Introduction* Oxford University Press Kindle Edition. (L 395)

[260] 2010, Rothery, David A. *Planets: A Very Short Introduction* Oxford University Press. Kindle Edition. (L 1120)

[261] 2007, Catling, David C. *Astrobiology: A Very Short Introduction* Oxford University Press Kindle Edition. (L 826)

[262] 2010, Rothery, David A. *Planets: A Very Short Introduction* Oxford University Press. Kindle Edition. (L 1144)

[263] 2007, Catling, David C. *Astrobiology: A Very Short Introduction* Oxford University Press Kindle Edition. (L 921)

[264] 2010, Rothery, David A. *Planets: A Very Short Introduction* Oxford University Press. Kindle Edition. (L 1097)

[265] 2007, Catling, David C. *Astrobiology: A Very Short Introduction* Oxford University Press Kindle Edition. (L 823)

[266] 2010, Rothery, David A. *Planets: A Very Short Introduction* Oxford University Press. Kindle Edition. (L 1143)

[267] 2010, Rothery, David A. *Planets: A Very Short Introduction* Oxford University Press. Kindle Edition. (L 1212)

[268] 2004, Ball, Philip, *The Elements: A Very Short Introduction* Oxford University Press. Kindle Edition. (L 727)

[269] 2010, Rothery, David A. *Planets: A Very Short Introduction* Oxford University Press. Kindle Edition. (L 666)

[270] 2007, Catling, David C. *Astrobiology: A Very Short Introduction* Oxford University Press. Kindle Edition. (L 233)

[271] 2010, Rothery, David A. *Planets: A Very Short Introduction* (Oxford University Press. Kindle Edition. L 1203)

[272] 2010, Rothery, David A. *Planets: A Very Short Introduction* Oxford University Press. Kindle Edition. (L 1084)

[273] 2014, Emspak, Jesse via Space.com; http://www.space.com/26447-mercury-composition-giant-impact.html

[274] 2008, Stannard, Russell, *Relativity: A Very Short Introduction* Oxford University Press. Kindle Edition. (L 743)

[275] 2010, Rothery, David A. *Planets: A Very Short Introduction* Oxford University Press. Kindle Edition. (L 774)

[276] 2010, Rothery, David A. *Planets: A Very Short Introduction* Oxford University Press. Kindle Edition. (L 774)

[277] 2010, Rothery, David A. *Planets: A Very Short Introduction* Oxford University Press. Kindle Edition. (L 873)

[278] 2010, Rothery, David A. *Planets: A Very Short Introduction* Oxford University Press. Kindle Edition. (L 1002)

[279] 2010, Rothery, David A. *Planets: A Very Short Introduction* Oxford University Press. Kindle Edition. (L 1175)

[280] 2007, Catling, David C. *Astrobiology: A Very Short Introduction* Oxford University Press. Kindle Edition. (L 1967)

[281] 2010, Rothery, David A. *Planets: A Very Short Introduction* Oxford University Press. Kindle Edition. (L 1149)

[282] 2004, Ball, Philip, *The Elements: A Very Short Introduction* Oxford University Press. Kindle Edition. (L 736)

[283] 2010, Rothery, David A. *Planets: A Very Short Introduction* Oxford University Press. Kindle Edition. (L 1120)

[284] 2010, Rothery, David A. *Planets: A Very Short Introduction* Oxford University Press. Kindle Edition. (L 1113)

[285] 2004, Ball, Philip, *The Elements: A Very Short Introduction* Oxford University Press. Kindle Edition. (L 734)

[286] 2010, Rothery, David A. *Planets: A Very Short Introduction* Oxford University Press. Kindle Edition. (L 1172)

[287] 2008, Seymour, Percy *Dark Matters: Unifying Matter, Dark Matter, Dark Energy, and the Universal Grid* (L 464) Kindle Edition.

[288] 2014, NASA/Goddard Space Flight Center via ScienceDaily.com; http://www.sciencedaily.com/releases/2014/09/140911180754.htm?utm_source=feedburner&utm_medium=email&utm_campaign=Feed%3A+sciencedaily%2Ftop_news%2Ftop_science+%28ScienceDaily%3A+Top+Science+News%29

[289] 2010, Rothery, David A. *Planets: A Very Short Introduction* Oxford University Press. Kindle Edition. (L 1219)

[290] 2010, Rothery, David A. *Planets: A Very Short Introduction* Oxford University Press. Kindle Edition. (L 1000)

[291] 2010, Rothery, David A. *Planets: A Very Short Introduction* Oxford University Press. Kindle Edition. (L 1183)

[292] 2010, Rothery, David A. *Planets: A Very Short Introduction* Oxford University Press. Kindle Edition. (L 1530)

[293] 2010, Rothery, David A. *Planets: A Very Short Introduction* Oxford University Press. Kindle Edition. (L 1191)

[294] 2007, Catling, David C. *Astrobiology: A Very Short Introduction* Oxford University Press, USA. Kindle Edition. (L1511)

[295] 2014, NASA/Goddard Space Flight Center via ScienceDaily.com; http://www.sciencedaily.com/releases/2014/06/140619205444.htm?utm_source=feedburner&utm_medium=email&utm_campaign=Feed%3A+sciencedaily%2Ftop_news%2Ftop_science+%28ScienceDaily%3A+Top+Science+News%29

[296] 2007, Catling, David C. *Astrobiology: A Very Short Introduction* Oxford University Press, USA. Kindle Edition. (L 392)

[297] 2011, Hall, Alan *Electron* (L 171) Kindle Edition.

[298] 2012, Blundell, Stephen J. *Magnetism: A Very Short Introduction* Oxford University Press. Kindle Edition. (L 1587)

[299] 2012, Tuniz, Claudio, *Radioactivity: A Very Short Introduction* Oxford University Press. Kindle Edition. (L 799)

[300] 2012, Blundell, Stephen J. *Magnetism: A Very Short Introduction* Oxford University Press. Kindle Edition. (L 1588)

[301] 2012, Blundell, Stephen J. *Magnetism: A Very Short Introduction* Oxford University Press. Kindle Edition. (L 1628)

[302] 2008, Seymour, Percy *Dark Matters: Unifying Matter, Dark Matter,Dark Energy, and the Universal Grid* (L1039) Kindle Edition.

[303] 2012, Blundell, Stephen J. *Magnetism: A Very Short Introduction* Oxford University Press. Kindle Edition. (L 1588)

[304] 2008, Seymour, Percy *Dark Matters: Unifying Matter, Dark Matter, Dark Energy, and the Universal Grid* (L 986) Kindle Edition.

[305] 2008, Seymour, Percy *Dark Matters: Unifying Matter, Dark Matter, Dark Energy, and the Universal Grid* (L 995) Kindle Edition.

[306] 2008, Seymour, Percy *Dark Matters: Unifying Matter, Dark Matter, Dark Energy, and the Universal Grid* (L 1085) Kindle Edition.

[307] 2012, King, Andrew, *Stars: A Very Short Introduction* Oxford University Press. Kindle Edition. (L 931) (p. 55)

[308] 2007, Akasofu, Syun-Ichi, *Exploring the Secrets of the Aurora* Kindle. (L 934)

[309] Europlanet Media Centre via ScienceDaily.com; http://www.sciencedaily.com/releases/2010/09/100924084612.htm

[310] 2007, Akasofu, Syun-Ichi, *Exploring the Secrets of the Aurora* Kindle. (L 931)

[311] Royal Astronomical Society (RAS) via ScienceDaily.com; http://www.sciencedaily.com/releases/2012/03/120327093612.htm

[312] University of Leicester via ScienceDaily.com; http://www.sciencedaily.com/releases/2010/08/100804080620.htm

[313] 2010, Europlanet Media Centre via ScienceDaily.com; http://www.sciencedaily.com/releases/2010/09/100924084612.htm

[314] 2010, ESA/Hubble Information Centre via ScienceDaily.com; http://www.sciencedaily.com/releases/2010/02/100211111537.htm

[315] 2010, NASA/Jet Propulsion Laboratory via ScienceDaily.com http://www.sciencedaily.com/releases/2010/09/100929191651.htm

[316] 2011, University College London via ScienceDaily.com; http://www.sciencedaily.com/releases/2011/04/110420143622.htm

[317] 2011, University College London via ScienceDaily.com; http://www.sciencedaily.com/releases/2011/04/110420143622.htm

[318] 2014, NASA; http://www.jpl.nasa.gov/video/?id=1277.

[319] 2014, NASA/Jet Propulsion Laboratory via ScienceDaily.com; http://www.sciencedaily.com/releases/2014/02/140211103826.htm

[320] 2016, NASA; http://www.nasa.gov/cassini and http://saturn.jpl.nasa.gov.
[321] 2010, Rothery, David A. *Planets: A Very Short Introduction* Oxford University Press. Kindle Edition. (L 1320)
[322] 2010, Rothery, David A. *Planets: A Very Short Introduction* Oxford University Press. Kindle Edition. (L 1283)
[323] 2010, Rothery, David A. *Planets: A Very Short Introduction* Oxford University Press. Kindle Edition. (L 1285)
[324] 2010, Rothery, David A. *Planets: A Very Short Introduction* Oxford University Press. Kindle Edition. (L 1338)
[325] 2007, Catling, David C. *Astrobiology: A Very Short Introduction* Oxford University Press Kindle Edition. (L 412)
[326] 2010, Rothery, David A. *Planets: A Very Short Introduction* Oxford University Press. Kindle Edition. (L 1328)
[327] 2016, Redd, Nola Taylor via Space.co; http://www.space.com/18563-pluto-temperature.html
[328] 2014, Sherlin Lecture, via Community College of Aurora; http://www.ccaurora.edu/sherlin-lecture-pluto-and-beyond
[329] 2014, NASA; http://science.nasa.gov/science-news/science-at-nasa/2014/14jul_pluto2015/

Summary

To summarize the Abstract

- *the auroral oval is a key mechanism that allows a solar system planet to electromagnetically interact with interplanetary space and with another planet from our solar system*

- *the electron is a universal quality essential to the electromagnetic movement and exchange of energy between planets*

- *electron encounters and astrological planetary energy transference may occur at the same time and place*

To summarize the Introduction

- *the auroral oval is one process of several that link the planets together*

- *the auroral oval is a very important process that leads to atomic reactions and modifications producing an effect on living matter here on Earth*

- *the electron is a universal quality that becomes the fundamental source of interaction from interplanetary space to the surface of a planet to interact with all life*

- *electrons are found on all planets in our solar system*

- *electrons pervade interplanetary space that surrounds the planets*

- *electrons are a self energizing transportation common quality*

- *electrons are a fundamental component for the assembly and disassembly of atoms within an electromagnetically conducting ionosphere especially when energized by a magnetosphere with an auroral oval*

To summarize The Electron

- *the electron has been studied and much good science has been performed by many highly qualified scientists*

- *electrons determine the color, reactivity, and shape of chemicals*

- *electrons function to contain nuclei and thus form atoms, molecules and matter*

- *the electromagnetic force that holds electrons in atoms and links atoms to one another to make molecules holds matter together*

- *electrons that are the farthest from the nucleus will determine the chemical interactions of the atom with other atoms*

- *a chemical reaction means that the electrons holding atoms together are rearranging themselves*

- *the electron is fundamental to atomic elements*

- *the electron is able to influence the space which surrounds it by its spin*

- *an increasing magnetic field allows energy to escape as the electron spins realign*

- *the electron as far as we know is itself fundamental*

- *electrons move in response to either an electric field or a temperature gradient*

- *electrons behave as much as waves as they do as particles*

- *the electron excites an atomic nucleus to life*

- *light can be emitted when an electron transfers from one orbit to another*

- *the electron has a magnetic field and magnetic poles with a north and south polarity*

- *the spinning direction of the electrons defines the direction of the magnetization in a material*

- *electrons reside in very specific orbitals, and that these orbitals have very specific energies*

To summarise The Auroral Oval

- *the actions of the electron and its manifestations within the ionosphere were well studied and explained by Professor Suyin-Ichi Akasofu*

- *at the rate of 1.5 billion volts an hour electrons spiral through the westward turning ring current that forms the mouth of the negatively charged 10.4 kilometer wide northern auroral oval that delineates Earth's polar cap*

- *the field-aligned currents flow between the magnetosphere and the ionosphere as a result of the magnetosphere-ionosphere coupling*

- *the upward field-aligned current in a sheet form, carried by the downward streaming electrons, causes the curtain form of an auroral arc*

- *the auroral zone is a circular belt in the geomagnetic coordinate system, centered around the geomagnetic pole*

- *auroral electrons penetrate into the polar upper atmosphere by moving along the outer boundary of the outer radiation belt*

- *the width of the auroral oval changes intermittently*

- *magnetospheric disturbances are various manifestations of the power generated by the solar wind-magnetosphere dynamo*

- *substorms are the cause of the ring current belt, injecting high-energy protons from the magnetotail into that belt*

- *the aurora can then be understood as the only visible manifestation of electrical discharge processes that are powered by the dynamo*

- *the ionosphere is an electromagnetic meeting place for highly energetic free electrons of varying strengths that split molecules apart and interact with neutral atoms and ions by changing their electric charge*

- *electrons in one atomic form or another continue downward passing out of the ionosphere*

- *as electrons cool they add to the electromagnetic field of Earth*

- *the electron reaches the troposphere where it becomes fully engaged with Earth's climatic weather patterns before touching land or water*

To summarise Planets and Interplanetary Space

- *on a continuous basis all the planets in our solar system are releasing electrons out into interplanetary space*

- *electrons are continuously arriving at planet Earth*

- *the solar wind plasma is the electron manifesting in one of its many forms*

- *with increasing distance from the Sun, the high-speed streams overtake the slower plasma, producing co-rotating interaction regions on their leading edges*

- *the interplanetary field assumes a spiral configuration*

- *the continuous high temperature of the Sun causes outward radiation that accelerates into cold interplanetary space*

- *the Dirac Sea is an invisible field in space that is filled with virtual electrons which can become real by the addition of energy*

- *a system that has only two energy levels is an electron spin in a magnetic field*

- *when the electron travels through cold interplanetary space it turns inward to create energy ready for release at a later time*

- *the electron internally creates energy while externally it creates a bed of electromagnetic energy*

- *changes in an electric field will produce changes in magnetic field, and vice versa, and a self-sustaining wave of varying electric and magnetic fields will propagate off into space*

- *magnetic fields influence the particle and magnetic environment of our Earth*

- *electrons interacting with the interplanetary magnetic field approach planet Earth*

- *electron controlled charged particles can be channelled along magnetic field lines towards the top of the atmosphere causing auroral glows*

- *each of the giant planets has a strong magnetic field*

- *the Earth's atmosphere differs from nearby planets in the complexity of its layering*

- *atmospheric molecules split by ultraviolet light can combine with others, by series of reactions described as photochemistry*

- *when a solar storm brings plasma from the Sun to the Earth, this distorts the Earth's magnetic field and causes unusual currents to flow in the ionosphere*

- *the tropospheric temperature of a planet is controlled by how effectively the lower atmosphere absorbs electromagnetic radiation*

- *the Earth's hydrologic cycle is the interplay between interior, surface, and atmosphere*

- *the Moon, Mercury, Venus and Mars have ionospheres*

- *Venus has a 2,000 kilometer diameter south polar vortex with clouds 60 kilometers above the surface that are drawn downwards by warmer temperatures*

- *convection occurs when air near the base of the troposphere is heated and expands making it buoyant, it will then rise, to be replaced by colder air displaced from above*

- eddy currents are induced by the conducting ionosphere of Venus

- other planets that have magnetic fields also have magnetospheres with similarities to Earth's

- Venus has an ionosphere, a layer of the atmosphere filled with charged particles

- inside the Sun ongoing fusion reaction melts matter into plasma liberating electrons

- particles in plasma on the Sun are charged and moving very fast and able to produce and interact with a magnetic field via magnetic field lines

- the solar wind is the continual, almost radial, out flowing of the solar corona

- magnetic lines of force diverge from coronal holes

- there is considerable evidence to show that the movements and positions of the planets do have an effect on the solar cycle and on violent events on the Sun

- the Hubble Space Telescope depicted a clear image of the auroral oval on Jupiter and Saturn

- variations in the aurora provide information on changes in the associated magnetosphere

- Saturn, like all magnetised planets, emits radio waves into space from the polar regions

- *on Saturn the radio emissions pulse while the auroras beat in tandem with the radio waves*

- *the Sun constantly emits particles that reach all the planets of the solar system as the solar wind*

- *when an electrically charged stream in interplanetary space gets close to a planet with a magnetic field, like Saturn or the Earth, the field traps the particles*

- *at Saturn auroral features can also be caused by electromagnetic waves generated when the planet's moons move through the plasma that fills Saturn's magnetosphere*

- *Enceladus is linked to Saturn by powerful electrical currents -- beams of electrons that flow back and forth between the planet and moon*

- *jets of gas and icy grains emanate from the south pole of Enceladus, which become electrically charged and form an ionosphere*

Professor Syun-Ichi Akasofu points the way forward when he makes the following statements;

"When one finds an interesting phenomenon, it is necessary to relate it to other significant phenomena and demonstrate that a new finding is worth paying attention to." [330]

"The new generation of scientists are encouraged to challenge the present paradigms and advance our understanding of electromagnetic phenomena around the Earth, in interplanetary space, and the heliosphere." [331]

"It is important for Solar Physicists, Interplanetary Physicists, Magnetospheric Physicists, and Upper Atmosphere Physicists to work together. There are many missing links among the four disciplines that will only be noticed if one attempts to synthesize space weather research." [332]

To this list of Solar Physicists, Interplanetary Physicists, Magnetospheric Physicists, and Upper Atmosphere Physicists it is necessary to add one more group of researchers, Astrologers. Astrologers have always said that the planets affect our lives here on Earth.

W.B.I. Beveridge in his The Art of Scientific Investigation says, "A scientific establishment is highly conservative and will attempt to preserve the power of its ruling group against any rebels. Thus, a pioneer often must stand-alone and be independent-minded on the fringe of the scientific establishment, and perhaps be a rebel." [333]

CPI Theory Part Two takes a step forward in showing that science has proven over and over again that planetary interaction is playing a role in how life unfolds here on Earth.

Sources for the Summary

[330] 2007, Akasofu, Syun-Ichi, *Exploring the Secrets of the Aurora* Kindle. (L 721)
[331] 2007, Akasofu, Syun-Ichi, *Exploring the Secrets of the Aurora* Kindle. (L 257)
[332] 2007, Akasofu, Syun-Ichi, *Exploring the Secrets of the Aurora* Kindle. (L 254)
[333] 2007, Akasofu, Syun-Ichi, *Exploring the Secrets of the Aurora* Kindle. (L 2376)

Conclusions for Part Two

Electromagnetic forces interact between the planets of our solar system. One method of electromagnetic interaction between a planet and interplanetary space is an auroral oval. A planet's ionosphere accepts the electromagnetic force facilitating reconstructive elemental changes that are disbursed downward through the atmospheric layers to act upon planetary life.

The electron and specifically its interaction with auroral ovals and ionospheres is a fundamental process for planetary interaction. All planets in our solar system possess an ionosphere and all planets in our solar system should have auroral ovals.

In order to have an ionosphere a planet must be magnetized. The evidence from scientific research suggests that all planets have ionospheres so it can be said that all planets are magnetized to a degree that corresponds to the strength of their ionosphere. The evidence from scientific research accepts that an auroral oval is caused by the combination of the solar wind and a magnetized planet. The visual effects of an auroral oval and the energy excitement of an ionosphere are caused by the electron. It seems a logical leap, that all planets in our solar system possess an ionosphere, are magnetized and should possess an auroral oval.

Continuous planetary interaction via electrons is a mechanism that should hold astrological signatures.

Acknowledgements for Part Two

Special thanks to Astrologer John Rutherford for the many hours of conversation on matters scientific. Always thanks to Astrologer Robert Currey for the many areas where he has been of great help. Thanks to Astrologer Bruce Scofield, PhD., for his advice. Thanks to Astrologer Robin Armstrong for the high energy input. Any errors are mine.

PART THREE

Abstract

CPI Theory Part Three advances the discussion of continuous planetary interaction within our solar system where the interaction is facilitated by electrons and the various electromagnetic states of the electron due to the changing temperature of the surrounding environment. CPI Theory Part Three will now explore why the electron behaves as it does and how its interaction could parallel astrological energy patterns and display astrological signatures. At all times I will attempt to stay focused on continuous planetary interaction.

Introduction

In CPI Theory Part One and CPI Theory Part Two it was said that aurora seen flowing from an auroral oval was the visualized radiation of the electromagnetic interaction between a planet and interplanetary space. CPI Theory further stated that a planet's ionosphere absorbs the electromagnetic interacting flow facilitating reconstructive essential changes. These essential changes are disbursed through the planetary atmospheric layers to act upon biological life. These essential changes also rise upwards and out through the auroral oval into interplanetary space.

Where do we start? There are two known types of magnetized planets in our solar system and both types can have an auroral oval and radiate aurora. Some planets like Earth have a magnetic core that generates a magnetic field. Other planets such as Neptune generate a magnetic field via the interaction of electrons within a highly gaseous atmosphere.

CPI Theory Part Three will explore continuous planetary interaction.

Magnetic Planets

All planets in our solar system have either a magnetic core or a magnetic field and sometimes both as in the case of Earth. It could be said that a magnetized core is magnetic to a degree corresponding to the strength of the ionosphere. And it could be said that the strength of a magnetic field will correspond to the size of the ionosphere.

CPI Theory says both a magnetic core and a magnetic field are part of the same mechanism namely electron interaction. Magnetic fields produce the magnetosphere. The interactions caused by the movement around a magnetized core, establishes an ionosphere. A magnetic field encapsulates an ionosphere.

All planets in our solar system have electromagnetically charged ionospheres. In order to have an ionosphere a planet must be magnetically active with either a magnetic core dynamo or a magnetic field. It is understood that to have an auroral oval a planet must have a magnetic field.

Professor Stephen J. Blundell says, "The Earth is a giant magnet. Our planet produces a magnetic field." [336]

Rodney A. Brooks, Doctor of Physics from Harvard University says, "...in the case of the Earth's magnetism, it is circulating electric currents in the Earth's core that create its magnetic field." [337]

Dr. Percy Seymour says, "The Earth's magnetic field is generated in the conducting material of its fluid outer core. Initially the dipole field will pass through this core, going from the north magnetic pole to the south magnetic pole. The outer core, which lies beneath the crust and the mantle, is rotating. However it does not rotate as a solid body. The rotation rate is latitude dependent – the material closer to the equatorial plane of Earth rotates faster than those regions closer to the poles. Thus, a single line of magnetic field, originally going from pole to pole, will be gradually distorted." [338]

One of the important properties of both solid and gaseous planets is their magnetic field and its magnitude. [339]

Professor David A. Rothery says, "In total, the Earth's core occupies about 16% of the planet's volume. Comparable values for Venus and Mars, which are estimates based largely on their average densities, are about 12% and 9%, respectively. There are some very limited seismic data from the Moon, hinting at a relatively small core between about 220 and 450 kilometres in radius (less than 4% of the Moon's total volume)." [340]

Independent researcher Nainan K. Varghese says, "It is generally believed that planetary magnetism is the result of the motion of molten iron alloys in a planetary body's core region. An assumed dynamo mechanism, in the core region of a planetary body, generates a magnetic field by a feedback loop... electric current loops in the region create a magnetic field and changing magnetic loops in turn create electric fields [341] ...Planetary magnetism is formed by planetary motion and free-floating atoms in relatively calm fluids near their surface. A solid core does not help and is not essential to form a magnetosphere about a planet." [342]

Earth has an inner core smaller than Earth's Moon. The inner core, once thought to be a solid ball of iron, has some complex structural properties. A team from the University of Illinois and colleagues at Nanjing University in China found a distinct inner-inner core, about half the diameter of the whole inner core. The iron crystals in the outer layer of the inner core are aligned directionally, north-south. However, in the inner-inner core, the iron crystals point roughly east-west. Not only are the iron crystals in the inner-inner core aligned differently, they behave differently from their counterparts in the outer-inner core. [343]

Jupiter and Saturn may not have much in the way of a rocky core but they do have a lot of hydrogen. It is thought that there is a considerable amount of hydrogen under high enough pressures that it enters a metallic phase. Electrical currents in this layer create strong magnetic fields. [344] Joseph Boyle at Quora.com says, "Under high enough pressure, hydrogen atoms are squeezed together enough so that electrons can move around instead of being trapped in diatomic molecules. This conductive state is known as metallic hydrogen." [345] Metallic hydrogen is a phase of hydrogen in which it behaves as an electrical conductor. [346]

Most of the interior of Jupiter is liquid and most of the liquid is primarily hydrogen. Jupiter has a small core that is perhaps a few tens of Earth masses. Under normal conditions hydrogen is not a metal and does not conduct heat or electricity very well. Scientists suggest that under the extreme pressure found deep inside Jupiter the electrons are released from the hydrogen molecules and are free to move about the interior. This causes hydrogen to behave as a metal becoming a conductor for both heat and electricity. The intense magnetic field of Jupiter is thought to result from electrical currents in this region of metallic hydrogen that is spinning rapidly and thought to compose 75% of the planet's mass. Jupiter radiates 1.6 times the energy than falls on it from the Sun meaning that

Jupiter should have an internal heat source. This internal heat source is presumably responsible for driving the complex weather pattern in Jupiter's atmosphere, unlike the Earth where the primary heat source driving the weather is the Sun. [347]

Uranus and Neptune have magnetic fields. Uranus is highly tilted with regards to its axis and is distinctively off-center. It is thought that their magnetic fields are generated by the movement of conductive fluids (mixtures of ammonia, methane and water) in a shell within the planet. [348]

Measurements made in 2013-2014 confirm the general trend that the Earth's magnetic field is weakening, with the most dramatic declines over the western hemisphere. But in other areas, such as the southern Indian Ocean, the magnetic field has strengthened. These changes are based on the magnetic signals stemming from Earth's core. There are magnetic contributions from other sources, namely the mantle, crust, oceans, ionosphere and magnetosphere. [349]

Professor Stephen J. Blundell says, "Every decade or so, compass needles in Africa are shifting by a degree, and the magnetic field overall on planet Earth is about 10% weaker than it was in the 19th century." [350]

Since the magnetic field of Earth changes over time then possibly the magnetic fields of other planets are also changing. From an astrological perspective any change of planetary magnetic strength could account for changes in the type of astrological influence. These magnetic changes speak to understanding why planets beyond Saturn and other solar system phenomena are considered in modern day astrological charts.

Astrologer John Rutherford says, "Astrology was more literal a century ago and has been made more subtle and psychologically focused since. This is likely induced by the use of electricity, pushing back the distance of the Earth's magnetic field, and weakening the intensity of responses to celestial changes. Also, the Sun and the planets appear as, and therefore are interpreted as, set or fixed objects in space. This is hardly the case. Energy from the Sun varies at times to an extreme degree, affecting the magnetic environment of all the planets. The best way to keep Astrological interpretations accurate is by timely observations, and adjusting interpretations as conditions change in our extended environment." [351]

In 2012 Professor Claudio Tuniz said, "...energy is dissipated, either gradually or abruptly, towards the external layers of Earth, but only a small fraction can be utilized. The amount of energy available depends on the Earth's geological dynamics, which regulates the transfer of heat to the surface of our planet. The total power dissipated by the Earth is 42 trillion watts: 8 trillion watts from the crust, 32.3 trillion watts from the mantle, 1.7 trillion watts from the core. This amount of power is small compared to the 174,000 trillion watts arriving to the Earth from the Sun." [352]

Sources for Magnetic Planets

[336] 2012, Blundell, Stephen J. *Magnetism: A Very Short Introduction* Oxford University Press. Kindle Edition. (L 1477)
[337] 2010, Brooks, Rodney A. *Fields of Color: The theory that escaped Einstein* Epic Publications. Kindle Edition. (L 824)
[338] 2008, Seymour, Percy *Dark Matters: Unifying Matter, Dark Matter, Dark Energy, and the Universal Grid* (L 643) Kindle Edition.

[339] 2014, Lomonosov Moscow State University
http://www.sciencedaily.com/releases/2014/11/141120141800.htm?
utm_source=feedburner&utm_medium=email&utm_campaign=Feed
%3A+sciencedaily%2Ftop_news%2Ftop_science+%28ScienceDaily
%3A+Top+Science+News%29
[340] 2010, Rothery, David A. *Planets: A Very Short Introduction*
Oxford University Press. Kindle Edition. (L 740)
[341] 2013, http://vixra.org/abs/1102.0038
[342] 2015, Varghese, Nainan K. March 22, 2015 via
www.Quera.com
[343] 2015, University of Illinois at Urbana-Champaign
http://www.sciencedaily.com/releases/2015/02/150209113222.htm
[344] 2015, Silver, Shereth via Quora.com March 19, 2015
[345] 2015, Boyle, Joseph via Quora.com March 20, 2015.
[346] 2016, http://en.wikipedia.org/wiki/Metallic_hydrogen
[347] http://csep10.phys.utk.edu/astr161/lect/jupiter/interior.html
[348] 2015, Silver, Shereth via Quora.com March 19, 2015
[349] 2014, European Space Agency
http://www.sciencedaily.com/releases/2014/06/140620115751.htm
[350] 2012, Blundell, Stephen J. *Magnetism: A Very Short
Introduction* Oxford University Press. Kindle Edition. (L 1520)
[351] 2015, Astrologer John Rutherford by permission.
[352] 2012, Tuniz, Claudio, *Radioactivity: A Very Short Introduction*
Oxford University Press. Kindle Edition. (L 605)

Planetary and Interplanetary Fields

In 1845 Michael Faraday introduced, as an explanation for electric and magnetic forces, the field concept where a field is a property or a condition of space. [353]

Dr. Rodney A. Brooks says, "...fields are a property of space, not a separate substance in space [354] ...Electric and magnetic fields have a real existence in themselves." [355]

Professor Steven Weinberg, Theoretical Physicist and Nobel laureate in Physics says, "Fields are conditions of space itself, considered apart from any matter that may be in it." [356]

Professor Frank Wilczek, Theoretical Physicist, Mathematician and a Nobel laureate says, "In quantum field theory, the primary elements of reality are not individual particles, but underlying fields. Thus all electrons are but excitations of an underlying field, the electron field, which fills all space and time." [357]

The electron, having a charge, produces an electromagnetic field around itself. In turn, this field, the so-called self-field of the electron, interacts with the electron. [358] The self-field of the electron is a surrounding electromagnetic field that is created by the electron. Even though the electron is localized it can travel and spread out over large distances. The electron is a quantum of the electron self field that affects the dynamics of the electron. [359] Art Hobson, Professor Emeritus of Physics University of Arkansas says, "An electron...is simply an energy increment of a spread-out matter field." [360]

Dr. Rodney A. Brooks says, "A quantum of the electromagnetic field can be emitted from an atom in the Sun, travel through millions of miles of space, spreading out as it goes, and then interact as a single unit with an atom in your eye [361] ...Quantum fields consist of a set of physical properties at every point of space, with equations that describe how these properties or field intensities influence each other and change with time [362] ...Fields can have mass and charge. And if everything is made of fields, then all physical properties – spin, mass, charge, energy, etc. – must be properties of fields [363] ...Quantum fields come in families - fields that have the same spin and obey similar equations." [364]

Physics Professor Robert L. Mills says, "The only way to have a consistent relativistic theory is to treat *all* the particles of nature as the quanta of fields, like photons. Electrons and positrons are to be treated as the quanta of the electron-positron field, whose classical field equation, the analog of Maxwell's equations for the electromagnetic field, turns out to be the Dirac equation, which started life as a relativistic version of the single-particle Schrödinger equation..." [365]

Dr. Rodney A. Brooks says, "The field intensity at a point can be a superposition of values [366] ...In quantum field theory, everything is fields...reality consists only of fields and interactions between fields..." [367]

Professor Steven Weinberg says, "Just as there is an electromagnetic field whose energy and momentum come in tiny bundles called photons, so there is an electron field whose energy and momentum and electric charge are found in the bundles we call electrons, and likewise for every species of elementary particles. The basic ingredients of nature are fields; particles are derivative phenomena [368] ...Furthermore, all these particles are bundles of the energy, or quanta, of various sorts of fields. A field like an electric or magnetic field is a sort of stress in space...The equations of a field theory like the Standard Model deal not with particles but with fields; the particles appear as manifestations of those fields." [369]

Dr. Rodney A. Brooks says, "Electromagnetism consists of two component fields, electric and magnetic. The electric involves static charges and the magnetic moving charges. Unlike gravity, which is always attractive, the electromagnetic force can be attractive, repulsive, or even sideways [370] ...The components that make up the electromagnetic field have directions associated with them [371] ...Electric and magnetic fields are two aspects of the way charged objects exert forces on each other [372] ...The electromagnetic field is a complex field that contains both electric and magnetic components that are really two aspects of the way electric charges interact. A charged particle at rest creates an electric field and another charged particle in its vicinity will feel a force because of the electric field. A charged particle in motion becomes an electric current and creates a magnetic field where a nearby moving charge will feel a force because of this magnetic field. It is for this reason that the two fields can be combined into a single electromagnetic field." [373]

Dr. Rodney A. Brooks also says, "The electromagnetic field contains energy, since it is made of electric and magnetic fields that exert forces. The stronger the field, the more pushing it can do and hence the more energy it contains [374] ...Electromagnetic fields created by different sources can combine to create a stronger or weaker field, depending on the direction of the forces. For example, two electric fields that push in opposite directions will result in a weaker field; if they are pushing in the same direction the force will be more intense [375] ...While the gravitational field is strongest near massive bodies and the electromagnetic field is strongest near charged bodies, both fields are found everywhere, even if their intensity is very weak [376] ...Like gravity the electromagnetic field is a force field, but unlike gravity it does not interact with everything. It is created by objects that carry an electric charge and it exerts a force on objects that carry an electric charge." [377]

Professor Stephen J. Blundell says, "James Clerk Maxwell realized that changes in an electric field will produce changes in a magnetic field, and vice versa, and a self-sustaining wave of varying electric and magnetic fields will propagate off into space." [378]

Using data from NASA's Kepler mission an international group of astronomers led by the University of Sydney has discovered strong magnetic fields are common in stars, not rare as previously thought. Associate Professor Dennis Stello says, "Because only five percent of stars were previously thought to host strong magnetic fields, current models of how stars evolve lack magnetic fields as a fundamental ingredient...Such fields have simply been regarded insignificant for our general understanding of stellar evolution...Our result clearly shows this assumption needs to be revisited...Now it is time for the theoreticians to investigate why these magnetic fields are so common." [379]

Sources for Planetary and Interplanetary Fields

[353] 2010, Brooks, Rodney A. *Fields of Color: The theory that escaped Einstein* Epic Publications. Kindle Edition. (L 245)
[354] 2010, Brooks, Rodney A. *Fields of Color: The theory that escaped Einstein* Epic Publications. Kindle Edition. (L 306)
[355] 2010, Brooks, Rodney A. *Fields of Color: The theory that escaped Einstein* Epic Publications. Kindle Edition. (L 251)

[356] 2001, Weinberg, S. *Facing Up: Science and its Cultural Adversaries* (Harvard University Press, Cambridge, MA.), p. 167: http://arxiv.org/ftp/arxiv/papers/1204/1204.4616.pdf

[357] 1999, Wilczek, F. *"Mass Without Mass I: Most of Matter,"* Physics Today 52 (11) http://arxiv.org/ftp/arxiv/papers/1204/1204.4616.pdf

[358] 1965, Tomonaga, S. (*Nobel Lecture 1965*) via 2010, Brooks, Rodney A. *Fields of Color: The theory that escaped Einstein* (L 2120) Epic Publications. Kindle Edition.

[359] 2010, Brooks, Rodney A. *Fields of Color: The theory that escaped Einstein* Epic Publications. Kindle Edition. (L2115)

[360] 2007, http://physics.uark.edu/hobson/pubs/07.02.TPT.pdf

[361] 2010, Brooks, Rodney A. *Fields of Color: The theory that escaped Einstein* Epic Publications. Kindle Edition. (L 269)

[362] 2010, Brooks, Rodney A. *Fields of Color: The theory that escaped Einstein* Epic Publications. Kindle Edition. (L 2278)

[363] 2010, Brooks, Rodney A. *Fields of Color: The theory that escaped Einstein* Epic Publications. Kindle Edition. (L 1337)

[364] 2010, Brooks, Rodney A. *Fields of Color: The theory that escaped Einstein* Epic Publications. Kindle Edition. (L 1553)

[365] 1994, Mills, R. *Space, Time, and Quanta: An Introduction to Modern Physics* (W. H.Freeman, New York), Chp 16: http://arxiv.org/ftp/arxiv/papers/1204/1204.4616.pdf

[366] 2010, Brooks, Rodney A. *Fields of Color: The theory that escaped Einstein* Epic Publications. Kindle Edition. (L 2296)

[367] 2010, Brooks, Rodney A. *Fields of Color: The theory that escaped Einstein* Epic Publications. Kindle Edition. (L 2339)

[368] 2001, Weinberg, S. *Facing Up: Science and its Cultural Adversaries* (Harvard University Press, Cambridge, MA.): http://arxiv.org/ftp/arxiv/papers/1204/1204.4616.pdf

[369] 1992, Weinberg, S. *Dreams of a Final Theory: The Search for the Fundamental Laws of Nature* (p. 25) Random House, Inc., New York: http://arxiv.org/ftp/arxiv/papers/1204/1204.4616.pdf

[370] 2010, Brooks, Rodney A. *Fields of Color: The theory that escaped Einstein* Epic Publications. Kindle Edition. (L 1173)

[371] 2010, Brooks, Rodney A. *Fields of Color: The theory that escaped Einstein* Epic Publications. Kindle Edition. (L 832)

[372] 2010, Brooks, Rodney A. *Fields of Color: The theory that escaped Einstein* Epic Publications. Kindle Edition. (L 826)

[373] 2010, Brooks, Rodney A. *Fields of Color: The theory that escaped Einstein* Epic Publications. Kindle Edition. (L 811)
[374] 2010, Brooks, Rodney A. *Fields of Color: The theory that escaped Einstein* Epic Publications. Kindle Edition. (L 1060)
[375] 2010, Brooks, Rodney A. *Fields of Color: The theory that escaped Einstein* Epic Publications. Kindle Edition. (L 841)
[376] 2010, Brooks, Rodney A. *Fields of Color: The theory that escaped Einstein* Epic Publications. Kindle Edition. (L 835)
[377] 2010, Brooks, Rodney A. *Fields of Color: The theory that escaped Einstein* Epic Publications. Kindle Edition. (L 803)
[378] 2012, Blundell, Stephen J. *Magnetism: A Very Short Introduction* Oxford University Press. Kindle Edition (L 775)
[379] 2016, University of Sydney
http://www.sciencedaily.com/releases/2016/01/160104130419.htm?utm_source=feedburner&utm_medium=email&utm_campaign=Feed%3A+sciencedaily%2Ftop_news%2Ftop_science+%28ScienceDaily%3A+Top+Science+News%29

Spin and Orbit

Spin

The spinning motion of the electron gives rise to a magnetic field. [380] Magnetism is caused by the motion of electrons, described by terms like orbital and spin. [381]

In an atom, magnetism arises from the spin and orbital momentum of its electrons. The electrons move in two ways: (1) spin, which can loosely be thought as spinning around themselves; (2) orbit, which refers to an electron's movement around the nucleus of its atom. The spin and orbital motion gives rise to the magnetization producing a magnetic field. The spinning direction of the electrons defines the direction of the magnetization in a material. [382]

Professor Richard P. Feynman said, "Dirac's theory said that an electron had a magnetic moment [383] ...When an electron responds to an external magnetic field it is called the magnetic moment." [384] Dr. Alan Hall says, "Electrons generate a magnetic moment because they spin and it is the direction of that spin that determines whether various electrons attract or repel each other." [385] Professor Peter Atkins says, "It is known that an increasing magnetic field allows energy to escape as the electron spins realign." [386]

Orbit

Professor Andrew King says, "The arrangement of the electron orbits is not arbitrary, but governed by definite physical rules.

Each orbit corresponds to a precise and distinct energy for the electron to occupy." [387]

Electrons can jump from an outer orbit to an inner orbit, but only to a limited extent to ensure the stability of the atom. The energy lost by the electron while jumping from the outer to the inner orbit is emitted as light by the atom. [388]

Professor Stephen J. Blundell says, "Quantum theory shows that the electrons orbits are not random, but are fixed in a small set of allowed orbits, each one of which is associated with a fixed value of energy. Light can be emitted when an electron transfers from one orbit to another. The energy of the emitted light makes up for the difference between the energies in the two orbits." [389]

Sources for Spin and Orbit

[380] 2010, Atkins, Peter *The Laws of Thermodynamics: A Very Short Introduction* Oxford University Press Kindle Edition. (L1447)
[381] 2010, Brooks, Rodney A. *Fields of Color: The theory that escaped Einstein* Epic Publications. Kindle Edition. (L 822)
[382] 2014, Ecole Polytechnique Fédérale de Lausanne http://www.sciencedaily.com/releases/2014/05/140508141829.htm?utm_source=feedburner&utm_medium=email&utm_campaign=Feed%3A+sciencedaily%2Ftop_news%2Ftop_science+%28ScienceDaily%3A+Top+Science+News%29
[383] 2014, Feynman, Richard P. *QED: The Strange Theory of Light and Matter* Princeton University Press. Kindle Edition. (L 317)
[384] 2014, Feynman, Richard P. *QED: The Strange Theory of Light and Matter* Princeton University Press Kindle Edition. (L1838)
[385] 2011, Hall, Alan *Electron* (L 281) Kindle Edition.
[386] 2010, Atkins, Peter *The Laws of Thermodynamics: A Very Short Introduction* Oxford University Press Kindle Edition. (L1458)

[387] 2012, King, Andrew, *Stars: A Very Short Introduction* (L 571) Oxford University Press. Kindle Edition.

[388] 2014, Sagar, Surendra Kumar, *Six Words* (L 637) Kindle Edition.

[389] 2012, Blundell, Stephen J. *Magnetism: A Very Short Introduction* Oxford University Press. Kindle Edition. (L 1107)

Waves

Dr. Rodney A. Brooks says, "An important property of waves is interference. Interference occurs when two fields converge at a point in space and either reinforce or cancel each other, depending on the direction of their forces [390] ...Electromagnetic waves are also more complicated than simpler water or air waves because the propagation involves an interplay between the electric and magnetic components, with changes in one inducing changes in the other." [391]

In quantum mechanics, particles do not have a distinct position in space. Instead, they exist as a wave function, a probability distribution that includes all the possible locations where a particle might be found. Electrons are elementary particles that are indivisible and unbreakable. New research suggests the electron's quantum state called the electron wave function can be separated into many parts. Experiments led by Humphrey Maris, Professor of Physics at Brown University, suggest that the quantum state of an electron known as the electron's wave function can be shattered into pieces and those pieces can be trapped in tiny bubbles of liquid helium. To be clear, the researchers are not saying that the electron can be broken apart. Electrons are elementary particles, indivisible and unbreakable. Professor Maris and his colleagues are suggesting that parts of the wave function can be separated and cordoned off from each other. [392]

Electrons from interplanetary space are continuously arriving at Earth where due to the electron's nature there can be many possibilities. The Schrödinger equation describes waves of possibility. [393] Professor Schrödinger developed an equation where the wave shape of an electron was defined. The equation revealed what happens inside an atom in a way that makes it possible to calculate the properties of more complex atoms and molecules. Physicist Paul Dirac said that the entire science of chemistry can be derived from the Schrödinger equation. [394]

In 1923, Physicist Louis de Broglie reasoned by analogy with Einstein's photon concept that if light which is clearly a wave can also be a particle, then an electron, which is clearly a particle can sometimes be a wave. In 1923 de Broglie deduced the wavelength of the electrons. [395] According to Physicist Niels Bohr a particle and a wave were complimentary concepts giving different representations of the same object. [396]

Astrophysicist John Gribbin says, "When an electron vibrates, it attempts to radiate by producing a field, which is a time symmetric mixture of a retarded wave propagating into the future and an advanced wave propagating into the past [397] ...If there is one particular link in the event chain that is special it is not the one that ends the chain. It is the link at the beginning of the chain when the emitter, having received various confirmation waves from its offer wave, reinforces one of them in such a way that it brings that particular confirmation wave into reality as a completed transaction." [398]

Physicist Louis De Broglie said, "A particle is a very small object, which is constantly localized in space and a wave is a physical process, which is propagated in space in the course of time...The wave has a very low amplitude and does not carry energy, at least not in a noticeable manner. The particle is a very small zone of highly concentrated energy incorporated in the wave, in which it constitutes a sort of generally mobile singularity. By reason of this incorporation of the particle in the wave, the particle possesses an internal vibration, which, as it moves, remains constantly in phase with the vibration of the wave..." [399]

Theoretical Physicist, Professor Richard P. Feynman said, "The electron does anything it likes. It just goes in any direction at any speed, forward or backward in time, however it likes, and when you add up the amplitudes it gives you the wave function." [400]

Two University of New Hampshire Space Physicists say that Kelvin-Helmholtz waves in Earth's atmosphere form when high-speed wind blows over more stagnant air masses creating turbulence that mixes the air masses. New research has shown that similar Kelvin-Helmholtz waves also frequently occur in Earth's magnetosphere and allow particles from the solar wind to enter the magnetosphere to produce oscillations that affect Earth's protective radiation belts. The underlying physical process that creates breaking wave cloud patterns in Earth's atmosphere also frequently opens the gates to high-energy solar wind plasma that perturbs Earth's magnetic field and magnetosphere. Joachim Raeder of the University of New Hampshire Space Science Center says, "Our paper shows that the waves, which are created by what's known as the Kelvin-Helmholtz instability, happens much more frequently than previously thought...this is significant because whenever the edge of Earth's magnetosphere, the magnetopause, gets rattled it will create waves that propagate everywhere in the magnetosphere, which in turn can energize or de-energize the particles in the radiation belts." Data from NASA's THEMIS mission found that Kelvin-Helmholtz waves occur at the magnetopause and can change the energy levels of our planet's radiation belts. Joachim Raeder says, "Previously, people thought Kelvin-Helmholtz waves at the magnetopause would be rare, but we found it happens all the time." [401]

As the steady solar wind rushes by Earth it creates Kelvin-Helmholtz waves. Kelvin-Helmholtz waves appear under a wide variety of conditions. It has been observed that just before the waves began, the plasmasphere a reservoir of charged gas around Earth sent out a thin plume of plasma that traveled over 20,000 miles to contact the edges of the magnetosphere, depositing additional atoms into that crucial Sun/Earth boundary. This case study suggests that the plume itself may trigger the waves. [402]

Chorus waves are intense whistler-mode electromagnetic emissions. The generation region of chorus waves is located outside the plasmapause near the geomagnetic equator where chorus waves are associated with enhanced fluxes of suprathermal electrons injected from the plasma sheet. Whistler-mode chorus waves have important roles in both acceleration and loss of energetic radiation belt electrons. Chorus wave activity is dependent on geomagnetic activity and occurs over a wide range of geospace. [403]

For the first time, scientists have observed ripples in the fabric of spacetime called gravitational waves, arriving at Earth. The signals came from two merging black holes 1.3 billion light years away. Gravitational waves carry information about their origins and are produced during the final fraction of a second of the merger of two black holes to produce a single, more massive spinning black hole. The gravitational waves were detected on September 14, 2015 at 5:51 a.m. Eastern Daylight Time (09:51 UTC) in Livingston, Louisiana, and Hanford, Washington, USA. According to general relativity when a pair of black holes orbit around each other they lose energy through the emission of gravitational waves. This causes the black holes to gradually approach each other over billions of years, and then much more quickly in the final minutes when the two black holes collide into each other at nearly one-half the speed of light and form a single more massive black hole, converting a portion of their combined mass to energy that is emitted as a final strong burst of gravitational waves. Gabriela González, Professor of Physics and Astronomy at Louisiana State University says, "This detection is the beginning of a new era: The field of gravitational wave astronomy is now a reality." [403a]

Astrologer Richard Tarnas, Professor of Psychology and Cultural History at the California Institute of Integral Studies in San Francisco says, "I believe that the evidence from both historical and biographical correlations suggests that the major aspects between planets should be seen as corresponding not to on-and-off light switches, as it were, with all the focus on the exact alignment within a degree or two, but rather as marking great archetypal wave forms that emerge, crest, and then surge through the collective or individual psyche and lifeworld." [404]

Sir Roger Penrose, a Mathematical Physicist said, "Probabilities do not arise at the minute quantum level of particles, atoms or molecules...but seemingly via some mysterious larger scale action connected with the emergence of a classical world that we can consciously perceive." [405]

Astrologer Isabel Hickey said, "It is not because you were born at a certain place or time that you react to influences, but the influences of that moment and that place in space show your potentials that can be actualized in the future." [405a]

Professor Lisa Randall is a Theoretical Physicist who studies Cosmology at Harvard University says, "On human time scales, the Universe and everything inside it is far from static. Not only do stars evolve, but galaxies do as well." [406]

Astrologer John Rutherford says, "Astrology is not a symbolic language in any static sense, but one deeply immersed in our dynamic world." [407]

Robin Armstrong, President of the RASA School of Astrology says, "To me there is an essential and even primal inner or spiritual (symbolic) nature to the language of astrology. It is the relationship between the symbolic and the spiritual that creates such an intriguing and fascinating relativity." [408]

Astrologer Isabel Hickey said, "Astrology deals with symbols, and the soul speaks and thinks in symbols." [408a]

Robin Armstrong, President of the RASA School of Astrology says, "Ancient Zoroastrian believed that when one died the soul returned to the Sun through the veins of the solar system and then were reborn through sunlight." [408b]

Astrologer Stephen Forrest says, "There is ultimately no metaphorical system so powerful and so primal as that of astrology." [409]

Astrologer Isabel Hickey said, "The horoscope is a blueprint of our character. Character IS destiny. There is nothing static in this universe in which we dwell. We can change by changing our attitudes and patterns of behavior. In so doing, we change our destiny." [409a]

Sources for Waves

[390] 2010, Brooks, Rodney A. *Fields of Color: The theory that escaped Einstein* Epic Publications. Kindle Edition. (L 1003)
[391] 2010, Brooks, Rodney A. *Fields of Color: The theory that escaped Einstein* Epic Publications. Kindle Edition. (L 903)

[392] 2014, Brown University
http://www.sciencedaily.com/releases/2014/10/141028214129.htm?
utm_source=feedburner&utm_medium=email&utm_campaign=Feed
%3A+sciencedaily%2Ftop_news%2Ftop_science+%28ScienceDaily
%3A+Top+Science+News%29
[393] 2002, Polkinghorne, John, *Quantum Theory: A Very Short
Introduction* Oxford University Press. Kindle Edition. (L 551)
[394] 2014, Sagar, Surendra Kumar, *Six Words* (L 670) Kindle
Edition.
[395] 2014, Sagar, Surendra Kumar, *Six Words* (L 666) Kindle
Edition.
[396] 2014, Sagar, Surendra Kumar, *Six Words* (L 920) Kindle
Edition.
[397] 2014, Sagar, Surendra Kumar, (extract from *"Q is for
Quantum"*) *Six Words* Kindle Edition (L 2000)
[398] 2014, Sagar, Surendra Kumar, *Six Words* (L 2280) (ref 6-009a)
Kindle Edition.
[399] 2014, Sagar, Surendra Kumar, *Six Words* (L 1991) Kindle.
[400] 2014, Sagar, Surendra Kumar, *Six Words* (L 1871) Kindle.
[401] 2015, University of New Hampshire
http://www.sciencedaily.com/releases/2015/05/150511125316.htm?
utm_source=feedburner&utm_medium=email&utm_campaign=Feed
%3A+sciencedaily%2Ftop_news%2Ftop_science+%28ScienceDaily
%3A+Top+Science+News%29
[402] 2015, NASA/Goddard Space Flight Center
http://www.sciencedaily.com/releases/2015/07/150708181714.htm
[403] 2009,
http://onlinelibrary.wiley.com/doi/10.1029/2009GL037595/full
[403a] 2016, Laser Interferometer Gravitational-wave Observatory
Laboratory
www.sciencedaily.com/releases/2016/02/160211103935.htm?utm_s
ource=feedburner&utm_medium=email&utm_campaign=Feed%3A+
sciencedaily%2Ftop_news%2Ftop_science+%28ScienceDaily%3A+
Top+Science+News%29
[404] 2015, Gryphon, Nina, NCGR Membership Director for the
NCGR Commentary March 24, 2015 cosmosandpsyche.com.
[405] 2014, Sagar, Surendra Kumar, (*The Emperor's New Mind*),
Six Words (L 2451) Kindle Edition.

[405a] 1992, 2011, Hickey, Isabel M. *Astrology, A Cosmic Science: The Classic Work on Spiritual Astrology* (L 142) SCB Distributors. Kindle Edition.

[406] 2015, Randall, Lisa, *Dark Matter and the Dinosaurs: The Astounding Interconnectedness of the Universe* Kindle Edition, Harper Collins. (L 1407)

[407] 2015, Astrologer John Rutherford by permission.

[408] 2015, Armstrong, Robin, President of the RASA School of Astrology by permission November 2015 http://www.rasa.ws/

[408a] 1992, 2011, Hickey, Isabel M. *Astrology, A Cosmic Science: The Classic Work on Spiritual Astrology* (L 24)   SCB Distributors. Kindle Edition.

[408b] 2015, Armstrong, Robin, President of the RASA School of Astrology by permission November 2015 http://www.rasa.ws/

[409] 2012, Forrest, Steven, *Yesterday's Sky: Astrology and Reincarnation* Seven Paws Press. Kindle Edition. (p. 32)

[409a] 1992, 2011, Hickey, Isabel M. *Astrology, A Cosmic Science: The Classic Work on Spiritual Astrology* (L 1502) SCB Distributors. Kindle Edition.

Electron Flow

When electrons exit a planet they rise out of the ionosphere passing through an auroral oval to become part of the electron sea that travels across interplanetary space. When electrons arrive at Earth's auroral oval they drop into the ionosphere and change places with other electrons that are either from Earth or from another astrological planet. Can an electron from another planet have signatures of that planet? I think so. Can an electron from Earth when it reaches another planet have signatures of Earth? I think so.

Our understanding of science is dictated by the state of the electron at a given moment. Take the electron out of any part of science and there is no science. The electron is necessary for everything. The electron is not only necessary but I think it has a purpose and destiny. Where you find energy interaction you will find the electron and should find astrological signatures. Robin Armstrong, President of the RASA School of Astrology says, "This statement has a potential for remarkable astrological and possible subjective insights." [411]

Electron flow is electrical current. There are two types of electron flow, direct current when electrons travel in only one direction and alternating current when the electrons flow in two directions, from positive to negative and from negative to positive, alternating between the two directions. Electric current occurs regardless of the direction of the electron flow. Dr. Alan Hall says, "Whenever electrons are disturbed we see the advent of electromagnetism." [412]

In conversation with Astrologer John Rutherford 2015:
TW...John, do you understand Earth's North Pole to be positive or negative?

JR...Neither. I don't see either pole as a positive or a negative they are just poles of an overall system.

TW...If Earth has a magnetized core then the North and South Poles are the poles of the core's dipole and should have a polarity, yes?

JR...Yes, but they can and do switch every so often. ...Whether a pole is said to be positive or negative is a relative thing, meaning it is only called positive or negative from conventional use. Just because you use your right hand doesn't make it right and the other hand wrong. What is so positive about a positive pole, that's only what it is called?

TW...Global magnetic flow diagrams clearly indicate inflow and outflow. The quote below says the Earth's North Pole is a negative pole. Are not positive ions attracted to a negative pole thus having the poles play a role in directing or attracting electron flow? [413]

Geophysical Methods in Geology by P.V. Sharma says, "By convention, the north-seeking pole corresponding to that at the north end of a compass needle is called the positive pole, and the south-seeking pole is referred to as the negative pole The lines of force are directed outward from a positive (i.e., north) pole and inward to a negative (i.e., south) pole." [414] Note that this means the Earth's north magnetic pole is a negative pole, because the positive north-seeking end of a compass needle is attracted toward it. [415]

Continued conversation with Astrologer John Rutherford 2015:
JR...Magnetic fields and electronic forces differ. The magnetic field curves round counter clockwise. The magnetic field lines of the Earth are static, meaning they don't move much but their presence curves the direction of electron flow. Electrons come into both poles they just curve in differing directions.

TW...Ok we are getting to where I want to go. I believe that there are three ways that electrons and particles are entering the ionosphere. (1) electrically (2) magnetically (3) physical pressure by solar wind that can be excessive with solar activity. [416]

A little confusion has been created by misnaming the electromagnetic flow of the electron. Electrons are said to have a negative charge because it was assumed electric charge moved in the opposite direction than it actually does. What was called negative actually had a surplus of electrons. When the true direction of electron flow was discovered positive and negative had already been so well established in the scientific community that no change was made. [417]

Lisa Murphy at the University of Illinois says that in semiconductors, when we say current flows from positive to negative, we don't mean the electron flow, we mean the hole flow. We know positive charges come from protons and negative charges come from electrons. But we also know that protons do not flow since the proton is stuck in the nucleus, therefore positive charges do not physically flow. Electrons on the other hand, do flow since they can jump from atom to atom so negative charges can flow. But when the negative charge jumps, it leaves behind a hole of positive charge that originates from the proton in the nucleus which no longer has its charge canceled to zero because the electron isn't there anymore. As the electrons move in one direction and leave behind holes, it will appear as though the holes flow in the opposite direction. This hole flow is conventional current. [418]

Electron flow can be quite clear if you just follow the electron and accept that the electron is guiding the flow and subsequent interaction.

Sources for Electron Flow

[411] 2015, Armstrong, Robin, President of the RASA School of Astrology by permission December 2015 http://www.rasa.ws/
[412] 2011, Hall, Alan *Electron* (L 301) Kindle Edition.
[413] 2015, Astrologer John Rutherford by permission October 2015
[414] 1986, Sharma, P. V. Geophysical Methods in Geology, 2nd Edition.
[415] 2015, http://geophysics.ou.edu/solid_earth/notes/mag_earth/earth.htm
[416] 2015, Astrologer John Rutherford by permission.
[417] 2015, http://www.allaboutcircuits.com/vol_1/chpt_1/7.html
[418] 2000, University of Illinois http://mste.illinois.edu/murphy/HoleFlow/HoleFlow.html

Ions, Ionization Energy and the Ionosphere

Ions

It is important to have a clear understanding of an ion. The simple answer is an ion is an atom with an electric charge.

Benjamin Lear, Professor of Chemistry and Shana Lear say, "Electrons determine whether or not the atom is an ion. An ion is an atom with an electrical charge, either positive or negative. This means that the number of protons (positive) and electrons (negative) are not equal...we will exclusively use the word ion to indicate an atom with a net electronic charge, and the word atom will imply an electronically neutral atom." [419]

The term ion was introduced by English Physicist and Chemist Michael Faraday in 1834. Ion was coined from the Greek word *ion* meaning to go or to walk, because ions move toward the opposite charge. [420]

An ion is an atom or molecule which has gained or lost one or more of its valence electrons, giving it a net positive or negative electrical charge. [421]

Astrologer John Rutherford says, "Every ion has a charge, either positive or negative. An ion whether positive or negative has electromagnetic components. A positive ion could have multiple protons, neutrons and electrons, but is still ionized if it is not neutral, meaning protons and electrons are balanced. A positive ion also has electrical and magnetic qualities, just opposite from a negative one." [422]

An ion is an atom or group of atoms in which the number of electrons is different from the number of protons. If the number of electrons is less than the number of protons, the particle is a positive ion, also called a cation. If the number of electrons is greater than the number of protons, the particle is a negative ion, also called an anion. [423]

An ion is an atom without electron/proton equilibrium. So it seems we are talking about an ion being an atom with an electric charge while an atom is an atom that is neutral and has no charge. Yes, terms like cation and anion add further complexity but do not forget ions are still atoms.

Based on the preceding definitions it can be seen that atoms come in three states negative, neutral and positive. Each state is dictated by the number of electrons. Compare the astrological qualities *cardinal, fixed* and *mutable* with the three types of atom; negative, neutral and positive. A negative atom displays electron arrival initiating via electron addition the *cardinal* quality. A neutral atom is at equilibrium and displays the *fixed* quality. A positive atom completes the process with the electron leaving and that displays the *mutable* quality leaving behind change.

From an astrological perspective I see a *cardinal* atom as *astrologically positive* and I see a *mutable* atom as *astrologically negative*. You add and that is positive thus *cardinal*. You take away and that is negative thus *mutable*. You neither add nor takeaway and that is neutral thus *fixed*. An entering electron is a *cardinal* electron. A vacating electron is a *mutable* electron. An electron that is not entering or vacating is a *fixed* electron.

When the electron is at equilibrium with the proton and not displaying an electromagnetic charge it is a neutral atom thus a *fixed* atom and if you like a *fixed* ion with no charge. When there is one more electron than proton the atom is a *cardinal* atom and a *cardinal* ion that is scientifically called a negative atom or a negative ion meaning there is an excess of the negative quality of the electron. And when there is one less electron than proton the atom is a *mutable* atom or *mutable* ion that is scientifically called a positive atom or positive ion meaning it has a positive amount of protons and the negative quality of the electron is reduced.

In the negative atom or negative ion state there is a positive amount of electrons meaning the electrons are on the excess side of equilibrium and thus *cardinal*. A *fixed* atom with a nucleus of protons and neutrons is surrounded by *fixed* electrons. This is a stable unit until a *cardinal* electron enters the atom by joining the *fixed* valence electrons. This electron addition causes one of the *fixed* valence electrons to vacate the atom thus becoming a *mutable* electron. This vacating *mutable* electron now seeks another atom to enter by becoming a *cardinal* electron that will cause another *mutable* electron to vacate a *fixed* atom. This is a continuous process where the activity level is governed by temperature. The higher the temperature the more active is the process of *mutable* electrons becoming *cardinal* electrons. *Mutable* electrons become *cardinal* electrons by adding one electron. A *mutable* electron realigns its spin when it pairs with another *mutable* electron and a *cardinal* electron is created.

Ionization Energy

Ionization is the electron in a transfer process.

Ionization energy is the quantity of energy that an atom must absorb to discharge an electron, resulting in a positive ion. Expelling the first electron will require less energy than expelling the second. Each successive electron requires more energy to be released. This is because after the first electron is lost, the overall charge of the atom becomes positive, and the negative forces of the electron will be attracted to the positive charge of the newly formed ion. The more electrons that are lost, the more positive this ion will be, the harder it is to separate the electrons from the atom. [424]

In general, the further away an electron is from the nucleus, the easier it is for it to be expelled. Ionization energy is a function of atomic radius. The larger the radius, the smaller the amount of energy required to remove the electron from the outer most orbital. Ionization energies are dependent upon the atomic radius. [425] Ionization energy is one of many properties of atoms which show marked periodicity. A pattern of overall increase in ionization energy recurs across all periods of the periodic table. [426]

The Ionosphere

The ionosphere is so named because it is the area where ions and ionization energy occur. When electromagnetically attracted electrons arrive at Earth's poles they enter the auroral oval and initiate changes within the ionosphere. Robin Armstrong, President of the RASA School of Astrology says, "This is an affirmative statement that could relate to astrological influence." [427]

Jeff Yee author of *The Particles of the Universe* says, "The electron is the key player within an ionosphere where the electrons create or disassemble life [428] ...Nuclear energy is created when molecule bonds are broken, but at a deeper level, when the parts of the atom's nucleus are broken and reassembled." [429] Not everyone thinks the same on this issue. Astrologer John Rutherford says, "Electrons do not create or disassemble life. It is a concert of all involved particles and molecules. Chemical energy is released not created, when molecular bonds are broken. Nuclear energy is released by fusion of light elements, or fission of heavy elements." [430]

Sources for Ions, Ionization Energy & the Ionosphere

[419] 2013, Lear, Benjamin; Lear, Shana, *The Atom* (L 64-65) Kindle.
[420] 2015, http://www.etymonline.com/index.php?term=ion
[421] 2015, http://chemistry.about.com/od/chemistryglossary/a/iondefinition.htm
[422] 2015, Astrologer John Rutherford by permission.
[423] 2015, http://whatis.techtarget.com/definition/ion
[424].2015, http://chemwiki.ucdavis.edu/Physical_Chemistry/Physical_Properties_of_Matter/Atomic_and_Molecular_Properties/Ionization_Energy
[425].2015, http://chemwiki.ucdavis.edu/Physical_Chemistry/Physical_Properties_of_Matter/Atomic_and_Molecular_Properties/Ionization_Energy
[426] 2011, Scerri, Eric R. *The Periodic Table: A Very Short Introduction* Oxford University Press. Kindle Edition. (L 1583)
[427] 2015, Armstrong, Robin, President of the RASA School of Astrology by permission December 2015 http://www.rasa.ws/
[428] 2012, Yee, Jeff, *The Particles of the Universe* (L 51) Kindle.
[429] 2012, Yee, Jeff, *The Particles of the Universe* (L 51) Kindle.
[430] 2015, Astrologer John Rutherford by permission.

Magnetotail

Professor Syun-Ichi Akasofu says, "The magnetotail is simply the tail of the ionosphere." [431] The magnetotail open region, fills with a building electron pressure that releases into the magnetotail to vector back to Earth's poles releasing electrons into the ionosphere.

Professor Akasofu continues, "Actually, a better parameter to examine changes of magnetic energy in the magnetotail is the size of the open region, which should be approximately proportional to magnetic flux in the magnetotail [432] ...The fact that the dimension of the open region is greater during an active period than during a quiet period shows that the magnetosphere is highly driven throughout substorm activity [433] ...The directly driven component is the one in which the energy derived from the solar wind is directly deposited in the magnetotail and the ionosphere..." [434]

Professor Akasofu also says, "Therefore, the decrease of the lobe-field need not be a result of magnetic reconnection and the subsequent transfer back of the tail flux as generally believed. Indeed, it makes perfect sense to consider that the decrease occurred because the e [electron energy] power, and subsequently the intensity of the solenoidal currents, began to decrease [435] ...One can reproduce qualitatively much of the magnetotail features by assuming that the intensity of the solenoidal current changes in harmony with e with a short time delay. Thus, it is quite natural for the tail field to decrease as e decreases [436] ...Considering the fact that the lobe-field can increase or decrease during substorms, it is reasonable to consider that the lobe-field increases when the e power is greater than the total loss rate (the energy dissipation rate) in the magnetosphere, while the lobe field decreases when the loss rate is greater than the *e*." [437]

Physics.org says, "The physical process that creates connections between the magnetic fields emanating from the Sun and a planet, a process known as magnetic reconnection, creates a portal through which solar plasma can penetrate the planetary magnetic field. The opening of these portals, known as flux transfer events, takes place roughly every 8 minutes at Earth and spawns a rope of streaming plasma." [438]

NASA is about to launch a mission called Magnetospheric Multiscale that consists of four spacecraft that will fly through Earth's magnetic field, to study reconnection in action. Dr. Jim Burch, Vice President of Space Science and Engineering at the Southwest Research Institute says, "A spectacular result of reconnection is known as the sawtooth crash. As heat in the tokamak builds up, the electron temperature reaches a peak then crashes to a lower value. Some of the hot plasma escapes. This is caused by reconnection of the containment field...In the expansive magnetic bubble that surrounds our planet, the process plays out over volumes as large as tens of kilometers across, for instance, when reconnection at the Sun propels plasma clouds toward Earth, where additional reconnection events then sparks auroras." [439]

Sources for Magnetotail

[431] 2007, Akasofu, Syun-Ichi, *Exploring the Secrets of the Aurora* Kindle. (L 1181)
[432] 2007, Akasofu, Syun-Ichi, *Exploring the Secrets of the Aurora* Kindle. (L 1192)
[433] 2007, Akasofu, Syun-Ichi, *Exploring the Secrets of the Aurora* Kindle. (L 1217)

[434] 2007, Akasofu, Syun-Ichi, *Exploring the Secrets of the Aurora* Kindle. (L 614)

[435] 2007, Akasofu, Syun-Ichi, *Exploring the Secrets of the Aurora* Kindle. (L 1222)

[436] 2007, Akasofu, Syun-Ichi, *Exploring the Secrets of the Aurora* Kindle. (L 1221)

[437] 2007, Akasofu, Syun-Ichi, *Exploring the Secrets of the Aurora* Kindle. (L 1224)

[438] 2012 http://phys.org/news/2012-11-high-frequency-flux-events-mercury.html#jCp

[439] 2015, NASA Science http://science.nasa.gov/science-news/science-at-nasa/2015/10mar_mms/

Atmosphere

Earth's atmosphere is the biochemical mixing ground located between our planet's surface and the outer edges of the magnetosphere. Encased within the magnetosphere, the atmosphere is constantly reacting to electromagnetic changes brought about by electrons entering the magnetosphere through the auroral oval. Meteor fragments and dust from interplanetary space striking the outer magnetosphere also affect the upper atmosphere.

Professor Lisa Randall says, "About 50 tons of extraterrestrial material enters the Earth's atmosphere every day, carried by millions of small meteoroids." [440]

Dr. Richard Collins, a Principal Investigator from the Geophysical Institute at the University of Alaska, Fairbanks, says, "Recent solar storms have resulted in major changes to the composition of the upper atmosphere above 49 miles, where enhancements in nitrogen compounds have been found. These compounds can be transported into the middle atmosphere where they can contribute to ozone destruction." [441]

The troposphere contains about 70% to 80% of the total mass of the Earth's atmosphere and 99% of the water vapor. Temperature and water vapor content in the troposphere decrease rapidly with altitude and thus most of the water vapour in the troposphere is concentrated in the lower, warmer zone. Water vapor concentrations are greatest above the tropics and decrease toward the polar regions. Winds increase with height and jet streams usually occur in the upper troposphere. The thickness of the troposphere varies from about 8 kilometers at the poles to about 18 kilometers at the

Equator. Increasingly, it is understood that air movements in the upper troposphere greatly influence weather systems in the lower troposphere. [442]

Cora Randall, a member of the AIM Science Team and the Chair of the Department of Atmospheric and Oceanic Sciences at the University of Colorado that monitors noctilucent clouds from Earth orbit says, "This past season [2014-2015] was not like the others...the clouds were much more variable, and there was an enormous decrease in cloud frequency 15 to 25 days after the summer solstice. That's when the clouds are usually most abundant... Preliminary indications are that it is indeed due to inter-hemispheric teleconnections...Previous research shows that noctilucent clouds are a sensitive indicator of long range teleconnectors in Earth's atmosphere, which link weather and climate across hemispheres." [443]

In January 2014 polar mesospheric cloud variations observed by AIM over Antarctica confirmed a link with planetary wave activity in the Arctic stratosphere. Planetary waves are also known as Rossby waves. The wintertime planetary wave activity in the stratosphere affects atmospheric circulation, leading to changes in both tropospheric weather in the winter hemisphere and mesospheric weather in the summer hemisphere. Previous observational studies of inter-hemispheric coupling have shown correlations between polar mesospheric clouds and wintertime stratospheric winds. This is the first observational study to show the link between polar mesospheric clouds and planetary wave activity, which is the driver of variability in the winds. It is also the first study to suggest a direct link between the polar summer mesosphere and the surface in the winter hemisphere. AIM satellite instruments have measured polar mesospheric clouds continuously with unprecedented precision since the mission was launched in 2007. During the summer polar mesospheric clouds occur 50 miles above the surface of the polar regions. [444]

Sources for Atmosphere

[440] 2015, Randall, Lisa, *Dark Matter and the Dinosaurs: The Astounding Interconnectedness of the Universe* Kindle Edition, Harper Collins. (L 1429)
[441] 2015, http://spaceweather.com/ January 26, 2015
[442] 2015, http://www.weatheronline.co.uk/reports/wxfacts/Troposphere.htm
[443] 2015, www.spaceweather.com March 1, 2015
[444] 2014, NASA http://aim.hamptonu.edu/mission/status_archive/status_20141013.php

Lightning

Lightning is another expression of the electron.

Lightning is found at Earth's exosphere and in Earth's atmosphere near ground level. Lightning is part of the process of a geomagnetic storm and a geomagnetic set of sub-storms that are caused by the increased injection pressure of the solar wind on the Earth's magnetosphere that leads to increased electron inflow through the auroral oval into the ionosphere. Temperature changes are also taking place causing thermoelectromagnetism. Lightning in the lower atmosphere is positive and negative interaction between water crystals in the air and the Earth caused by temperature changes, the same activity that facilitates electrons. [445]

The physicsclassroom.com says it best. At the beginning of any lightning strike the tops of the storm clouds acquire an excess of positive charge and the bottoms of the storm clouds acquire an excess of negative charge. In the lower atmosphere lightning involves a separation of charge. Clouds contain countless millions of suspended water droplets and ice particles moving in turbulent fashion. Additional water from the ground evaporates and rises upward to collide with water droplets within the clouds. In collisions electrons are ripped off the rising droplets, causing a separation of negative electrons from a positively charged water droplet or a cluster of droplets. [446]

In the upper atmosphere lightning involves a freezing process. Rising moisture encounters cooler temperatures at higher altitudes. These cooler temperatures cause the cluster of water droplets to undergo freezing. The frozen portion of the cluster of rising moisture becomes negatively charged and the outer droplets acquire a positive charge. Air currents within the clouds can rip the outer portions off the clusters and carry them upward toward the top of the clouds. The frozen portion of the droplets with their negative charge tends to gravitate towards the bottom of the storm clouds. In the end, a storm cloud becomes polarized with positive charges carried to the upper portions of the clouds and negative portions gravitating towards the bottom of the clouds. The cloud's electric field stretches through the space surrounding it and induces movement of electrons upon Earth where electrons on Earth's outer surface are repelled by the negatively charged cloud's bottom surface. This creates an opposite charge on the Earth's surface where buildings, trees and people can experience a buildup of static charge as electrons are repelled by the cloud's bottom. [446a]

A lightning bolt begins with the development of a step leader. Excess electrons on the bottom of the cloud begin a journey through the conducting air to the ground at speeds up to 60 miles per second. These electrons follow zigzag paths towards the ground, branching at various locations. As the step leader grows it provides the roadway between cloud and Earth along which the lightning bolt will eventually travel. As the electrons of the step leader approach the Earth, there is an additional repulsion of electrons downward from Earth's surface. The quantity of positive charge residing on the Earth's surface becomes even greater. This charge begins to migrate upward through buildings, trees and people into the air. This upward rising positive charge, known as a streamer, approaches the step leader in the air above the surface of the Earth at an altitude equivalent to the length of a football field. Once contact is made between the streamer and the leader, a complete conducting pathway is mapped out and the lightning begins. The contact point between ground charge and cloud charge rapidly ascends upward. As many as a billion trillion electrons can transverse this path in less than a millisecond. This initial strike is followed by several secondary strikes or charge surges in rapid succession. These secondary surges are spaced apart so closely in time that may appear as a single strike. The enormous and rapid flow of charge along this pathway between the cloud and Earth heats the surrounding air, causing it to expand violently. The expansion of the air creates a shockwave that we observe as thunder. [447]

University of Alabama Huntsville researchers found that wherever you are, if it's 8 a.m., it's time for the most powerful lightning strikes of the day. Not the largest number of lightning strikes, just the most powerful. The scientists found a cycle that seems to apply everywhere, although the difference between relative strengths of the morning peak and the afternoon lull can vary significantly from one region to the next. The peak power spike, however, is constant despite regional daily variations in lightning frequency. [448] Dr. Maher A. Dayeh, a Research Scientist with South Western Research Institute, Space Science and Engineering Division says, "Lightning strikes the Earth more than 4 million times a day..." [449]

Robin Armstrong, President of the RASA School of Astrology says, "Looking into the Gauquelin sectors for the 12th house would seem in order, as the planets, Moon and Sun across a day actually travel from the IC (Imum Coeli) to the 3rd house, to the 2nd house, to the 1st house reaching the Ascendant and then entering the 12th house, at the end of which the area of peak lightning power occurs. Michel Gauquelin found that in the charts of athletes Mars was often in the 12th house. This takes us into the validity of diurnal houses. We see it very clearly if we go into hemispheric assessments of rising and setting planets, or whether they are above or below the horizon." [450]

Researchers at University of Reading (UK) found a link between increased thunderstorm activity on Earth and streams of high-energy particles accelerated by the solar wind, offering compelling evidence that particles from interplanetary space help trigger lightning bolts. The researchers propose that the electrical properties of the air are somehow altered as the incoming charged particles from the solar wind collide with the atmosphere. [451] Researchers found that over a five year period (2001-2006) the UK experienced around 50% more lightning strikes when the Earth's magnetic field was bent by the Sun's own magnetic field. Dr. Matt Owens says, "The Sun's magnetic field is like a bar magnet, so as the Sun rotates its magnetic field alternately points toward and away from the Earth, pulling the Earth's own magnetic field one way and then another." [452]

An electron emits light as a by-product of one of its transfer processes. Lightning is one form of the electron facilitating transfer. No electron then no lightning. In the electromagnetic world science says there is a constant desire for equilibrium, the seeking of balance between negative and positive charges. Lightning is a thin elongated connection that emits light while connecting atmospheric electrons to the ground and deeper into Earth's core regions. [453]

Sources for Lightning

[445] 2014, Institute of Physics
http://www.sciencedaily.com/releases/2014/11/141119204849.htm?
utm_source=feedburner&utm_medium=email&utm_campaign=Feed

%3A+sciencedaily%2Ftop_news%2Ftop_science+%28ScienceDaily%3A+Top+Science+News%29

[446] 2015, http://www.physicsclassroom.com/class/estatics/Lesson-4/Lightning

[446a] 2015,
http://www.physicsclassroom.com/class/estatics/Lesson-4/Lightning

[447] 2015, http://www.physicsclassroom.com/class/estatics/Lesson-4/Lightning

[448] 2015, University of Alabama Huntsville
http://www.sciencedaily.com/releases/2015/03/150316185442.htm

[449] 2015, Southwest Research Institute
http://www.sciencedaily.com/releases/2015/05/150505121328.htm?utm_source=feedburner&utm_medium=email&utm_campaign=Feed%3A+sciencedaily%2Ftop_news%2Ftop_science+%28ScienceDaily%3A+Top+Science+News%29

[450] 2015, Armstrong, Robin, President of the RASA School of Astrology by permission November 2015 http://www.rasa.ws/

[451] 2014, Institute of Physics
http://www.sciencedaily.com/releases/2014/05/140514205758.htm

[452] 2014, Institute of Physics
http://www.sciencedaily.com/releases/2014/11/141119204849.htm?utm_source=feedburner&utm_medium=email&utm_campaign=Feed%3A+sciencedaily%2Ftop_news%2Ftop_science+%28ScienceDaily%3A+Top+Science+News%29

[453] 2014, Institute of Physics
http://www.sciencedaily.com/releases/2014/11/141119204849.htm?utm_source=feedburner&utm_medium=email&utm_campaign=Feed%3A+sciencedaily%2Ftop_news%2Ftop_science+%28ScienceDaily%3A+Top+Science+News%29

Temperature and Superconductivity

Temperature

Electrons move in response to either an electric field or a temperature gradient. [454]

Professor Frank Close says, "As electrons in an atom encounter increasing temperature they are raised to ever higher energy levels, until eventually they are ejected from the atom leaving the atom ionized." [455]

An atom is a nucleus of protons and neutrons surrounded by electrons. This is a stable unit until and new electron enters the atom by joining with the valence electrons. This causes one of the original valence electrons to vacate the atom. This vacating electron now seeks another atom to enter thus causing another electron to vacate the atom. This is a continuous process where the activity level is governed by temperature. The higher the temperature the more active are the entering electrons. [456]

Professor Stephen J. Blundell says, "It was already well known that the resistance of a metal fell as you cooled it. We now understand this as being due to the reduction of vibrations of the atoms in a solid that accompany cooling. The atoms in a solid wobble around like a vibrating jelly, but there is less jiggling around at colder temperatures. The electrons move through the solid when you try to pass an electrical current, but electrons can be deflected when they interact with vibrating atoms, and this deflection is called scattering. At low temperature, fewer vibrations mean less scattering of electrons and the resistance falls." [457]

180

Superconductivity

Professor Stephen J. Blundell says, "The Theory of Superconductivity assumes that there is attractive interaction of electrons [458] ...The idea is that the state of matter is protected by some collective quantum behavior, so that the microscopic constituents of the system act jointly together [459] ...For superconductivity to occur, the material has to be cooled to a very low temperature..." [460]

Richard Newrock, Professor of Physics at the University of Cincinnati, says that Professor Brian Josephson predicted in 1962 that pairs of superconducting electrons could tunnel right through the nonsuperconducting barrier from one superconductor to another. He also predicted the exact form of the current and voltage relations for the junction. Professor Josephson was awarded the 1973 Nobel Prize in Physics for his work. What occurs is that the electrons in the metal become paired. Above the critical temperature, the net interaction between two electrons is repulsive. Below the critical temperature, though, the overall interaction between two electrons becomes very slightly attractive. This very slight attraction allows them to drop into a lower energy state, opening up an energy gap. Because of the energy gap and the lower energy state, electrons can move and therefore current can flow. There is no electrical resistance in a superconductor, and therefore no energy loss. There is, however, a maximum supercurrent that can flow, called the critical current. Above this critical current the material is normal. When a metal goes into the superconducting state, it expels all magnetic fields, as long as the magnetic fields are not too large. [461]

Professor Stephen J. Blundell says, "Giaever's superconducting tunnel junction consisted of two superconductors separated by a very thin insulating layer. By making an electrical circuit with this tunnel junction it was possible to see how electrons could flow through the insulating layer, an effect which is impossible in classical physics. This is accomplished because of the ability of electrons to perform a ghost-like process of tunnelling through the insulating barrier [462] ...superconducting currents can flow across a Josephson junction and because these are superconducting currents, there is no dissipation of energy; hence, by applying a voltage one is giving the electrons energy that they do not need and cannot easily dissipate; the result is that the electrons oscillate back and forth across the junction, radiating energy as electromagnetic waves." [463]

Sources for Temperature and Superconductivity

[454] 2014, Massachusetts Institute of Technology
http://www.sciencedaily.com/releases/2014/07/140728104714.htm?utm_source=feedburner&utm_medium=email&utm_campaign=Feed%3A+sciencedaily%2Ftop_news%2Ftop_science+%28ScienceDaily%3A+Top+Science+News%29
[455] 2004, Close, Frank, *Particle Physics: A Very Short Introduction* Oxford University Press. Kindle Edition. (L 688)
[456] 2015, Northwestern University
http://www.qrg.northwestern.edu/projects/vss/docs/power/2-whats-electron-flow.html
[457] 2009, Blundell, Stephen J. *Superconductivity: A Very Short Introduction* Oxford University Press. Kindle Edition. (L 608)
[458] 2009, Blundell, Stephen J. *Superconductivity: A Very Short Introduction* Oxford University Press. Kindle Edition (L 1151)
[459] 2009, Blundell, Stephen J. *Superconductivity: A Very Short Introduction* Oxford University Press. Kindle Edition (L 2107)
 [460] 2009, Blundell, Stephen J. *Superconductivity: A Very Short Introduction* Oxford University Press. Kindle Edition. (L 356)
[461] Scientific American

http://www.scientificamerican.com/article/what-are-josephson-juncti/

[462] 2009, Blundell, Stephen J. *Superconductivity: A Very Short Introduction* Oxford University Press. Kindle Edition. (L 1389)

[463] 2009, Blundell, *Stephen J. Superconductivity: A Very Short Introduction* Oxford University Press. Kindle Edition. (L 1423)

Aurora

Science has stated that an auroral oval is caused by the combination of the solar wind and a planet with a magnetic field. The visual effects of an auroral oval and the energy excitement within an ionosphere are caused by electrons. Aurora can come in different colors due to the atmospheric particles interacting in the ionosphere.

Astrologer John Rutherford says, "The aurora colour is not from an excess of any element in the atmosphere. The colours relate to the degree of ionization of the elements available, i.e. the strength of the storm. The purples are mostly seen closer to the Arctic where the strength of the storm is most focused. It is also the strength of the storm that can ionize the upper atmosphere far into lower latitudes." [465]

Carla Helfferich, writing for the Alaska Science Forum, says that on the Sun, between March 12th-13th, 1989, enormous flares erupted extending 70,000 miles into interplanetary space. Electrons and protons were cast off by the Sun to stream outward into interplanetary space as an enhanced flow called the solar wind. When the solar wind and the Earth's magnetic field interact, electrical power is produced. The resulting electrical discharge in the upper atmosphere gives off photons that we see as aurora. In a typical auroral display, the light is a mixture of many colors. Green auroras, and green auroras with a reddish lower border, occur at an altitude near 100 kilometers above the Earth. Rare, all-red auroras occur much higher, at 300 to 500 kilometers altitude and are associated with a large influx of electrons. Red aurora is associated with intense solar activity and heating of the upper atmosphere from a large influx of low-energy electrons. These low energy electrons are moving too slowly to penetrate deeply into the atmosphere. [466]

Black aurora occurs where there are holes in the ionosphere where particles are shooting upwards into interplanetary space inside regions known as positively charged electric potential structures. This is opposite to the process that creates visible aurora, where electrons spiral down from interplanetary space into the atmosphere. Professor Göran Marklund of the Alfvén Laboratory in Sweden says, "The black aurora isn't actually an aurora at all; it's a lack of auroral activity in a region where electrons are sucked from the ionosphere..." [467]

Tomas Karlsson a Researcher in Space and Plasma Physics at Sweden's KTH Royal Institute of Technology and Dr. Andrew Wright, a Researcher at the School of Mathematics and Statistics in University of St Andrews say, "The depleted density and electrical conductivity in a black aurora substantially modify the wave reflected from the ionosphere, producing signatures in the magnetosphere...Electron currents follow the lines of magnetic force generated by the Earth's core and flow through the magnetosphere -- When these electrons meet with atoms of oxygen and nitrogen about 90 to 300 kilometers above Earth, they form colourful beams. Researchers found that inside these colourful displays, another process is happening that creates the black regions. Ordinary aurora are associated with downward-flowing electrons bombarding the atmosphere, the black ones are associated with electrons being sucked out from the atmospheres into interplanetary space leaving deep cavities in the upper ionosphere." According to the Researchers' model, if the downward current intensifies, it can cause a large number of electrons to move upward into the magnetosphere, thus depleting the ionosphere and creating a density cavity. [468]

A drop in low-energy electrons, long thought to have little or no effect corresponds with especially fast changes in the shape and structure of pulsating auroras. Though all auroras are caused by energetic particles, the source is different for pulsating auroras and active auroras. Active auroras happen when a dense wave of solar material, a high-speed stream or a coronal mass ejection, hits Earth's magnetic field releasing electrons in the magnetotail. Once released, electrons race towards the poles, interact with particles in Earth's upper atmosphere to create glowing lights that stretch across the sky in long ropes. Pulsating auroral electrons are sent spinning to the surface by complicated wave motions in the magnetosphere. These wave motions can happen at any time, not just when a wave of solar material rattles the magnetic field. Only after a collision from the first set of higher-energy electrons, some of the secondary electrons shoot back upwards along the magnetic field line, and towards the opposite hemisphere. [469]

According to new observations from NASA immense cracks sometimes develop in Earth's magnetosphere and remain open for hours. This allows the solar wind to gush through and power stormy space weather. New observations revealed an area almost the size of California in the arctic upper atmosphere where a 75 megawatt proton aurora flared for hours. A proton aurora is a form of aurora caused by heavy solar ions striking Earth's upper atmosphere, causing it to emit ultraviolet light. IMAGE was able to view the proton aurora for more than 9 hours implying that the crack remained continuously open. Researchers estimate that the crack was twice the size of Earth at the boundary of our magnetic shield that is located about 38,000 miles above the planet's surface. Since the magnetic field converges, as it enters the Earth in the polar regions the crack narrowed to about the size of California down near the upper atmosphere. [470]

The zenith is an area in the astrological chart where the auroral oval empties its electrons due to solar pressure and also the area in the astrological chart where this influence can be witnessed to occur. It could be said that the zenith is the center of the inflow radiation into the zodiacal chart while the eastern horizon is the entrance into the zodiacal chart and into the natural world where we live. These combined energies form a horizontal plane spreading from the eastern horizon to the western horizon representing the auroral oval's electromagnetic interaction of the planets within a zodiacal chart.

The hemispheric power is an energy estimate of all electrons precipitating into a hemisphere. [471] The electron hemispheric power is found in both hemispheres on an hourly basis. The sum of the northern and southern hemispheres is the auroral electron power. [472]

There is solar forcing from the interplanetary magnetic field and solar wind velocity on particles at the auroral inputs. [473] Hemispheric power levels are estimated from observations made of high-velocity electrons and positive ions raining down over the auroral oval. All of the energy, of these particles is eventually deposited into the Earth's upper atmosphere. However, the energy of these particles is only about 10% to 30% of the total energy flowing into the atmosphere along the auroral oval. The remainder is deposited by up to million-ampere electric currents as they heat the upper atmosphere. During times of high activity, the power deposited can easily exceed the sum of all the electrical power generated in the United States over the same time interval. [474]

A 2016 Discover magazine article says, "For the first time, astronomers have spotted an aurora, akin to our northern and southern lights, shimmering on a world outside our solar system. The find may bolster the search for extraterrestrial life, since the magnetic fields that drive auroras likely keep planets habitable. Every planet in our solar system (and even some moons) with a moderate magnetic field boasts these celestial light shows. They occur when charged space particles, typically from the Sun, stream along a planet's magnetic field lines and interact with atmospheric atoms, producing not only optical light but also radio emissions. Dr. Gregg Hallinan an Assistant Professor of Astronomy at the California Institute of Technology and colleagues have detected both types of radiation from what appears to be a brown dwarf, an object that straddles the boundary between planet and star. The world's aurora, reported in *Nature* in July as about a million times brighter than Earth's, suggests that brown dwarfs have magnetic activity more like planets than stars. Dr. Hallinan hopes more observations shed light on the origins of the charged particles that power the aurora, which are currently unknown." [475]

Sources for Aurora

[465] 2015, Astrologer John Rutherford by permission October 2015
[466] 1989, Helfferich, Carla, *The Rare Red Aurora* Article #918, University of Alaska, March 22, 1989
http://www2.gi.alaska.edu/ScienceForum/ASF9/918.html
[467] 2001, European Space Agency http://sci.esa.int/cluster/29100-cluster-quartet-probes-the-secrets-of-the-black-aurora/
[468] 2015, KTH, The Royal Institute of Technology
http://www.sciencedaily.com/releases/2015/04/150415090904.htm
[469] 2015, NASA/Goddard Space Flight Center
http://www.sciencedaily.com/releases/2015/10/151007185043.htm

[470] 2003, NASA Science http://science.nasa.gov/science-news/science-at-nasa/2003/03dec_magneticcracks/

[471] 2014, http://cedarweb.hao.ucar.edu/wiki/index.php/DMSP:ssj4_hp

[472] 2009, University Corporation for Atmospheric Research http://www.acd.ucar.edu/Events/Meetings/HEPPA/pdf_files/Sources_Transport/Emery.pdf

[473] 2009, University Corporation for Atmospheric Research http://www.acd.ucar.edu/Events/Meetings/HEPPA/pdf_files/Sources_Transport/Emery.pdf

[474] 2016, SpaceWeather https://www.windows2universe.org/spaceweather/quicklook4g.html

[475] 2016, January/February Discover Magazine http://discovermagazine.com/2016/janfeb/64-alien-aurora

The Sun and the Heliosphere

The Sun

The primary science goals for the 2018 Solar Probe Plus mission are to trace the flow of energy and understand the heating of the solar corona and to explore the physical mechanisms that accelerate the solar wind and energetic particles. [476] There is still a lot of uncertainty as to how the Sun affects the climate, but a study from Lund University in Sweden suggests that direct solar energy is not the most important factor, but rather indirect effects on atmospheric circulation. [477]

Solar maximum is a time of action, marked by massive explosions and dangerous space weather, while solar minimum is a time of quiet, when almost nothing happens. Solar activity never stops, it just changes form. Each extreme stirs up a unique mix of space weather that affects different parts of the Earth/Sun environment in varied and complex ways. Solar minimum and maximum are opposite extremes of the solar rhythm. [478]

Astrology Researcher Brian Johnson says, "Out of hundreds of observations over the last year what has been seen is that the coronal mass ejections and flares are closely timed to coincide with the timing of the planetary aspects as viewed from the Earth...when the inner planets aspect the outer planets there are massive storms on the Sun...." [479] Robin Armstrong, President of the RASA School of Astrology says, "Both Theodore Landscheit and Commander Nelson in their material on sunspots have mentioned this." [480]

Dr. Percy Seymour says, "Activity on the Sun can modulate the charged cosmic ray particles that come from beyond the solar system. One well-known way in which this can occur is called the Forbush decrease. Professors Thomas Gold and Eugene Parker both proposed theories to account for this sudden decrease in extrasolar system cosmic ray flux, which follows a large flare event on the Sun. According to Gold, a plasma cloud ejected by a large flare event can pull out the solar magnetic field to form a large magnetic bottle...that can eventually engulf the Earth and thus provide an additional shield from cosmic ray particles. Parker suggests that a large flare event gives rise to a blast wave (similar to a sonic boom), in which the strength of the interplanetary field is increased. The blast wave propagates out into space, and when it reaches the Earth it provides additional protection against cosmic rays. The abrupt increase in the interplanetary magnetic field within a blast wave has been observed by some spacecraft, thus providing support for Parker's theory." [481]

DSCOVR left Earth February 11, 2015 where it will monitor the solar wind one million miles upstream from Earth. DSCOVR will replace ACE as our primary warning system for solar magnetic storms. In addition to monitoring the solar wind, DSCOVR will also look back at Earth. [482]

The Heliosphere

Southwest Research Institute says, "The heliosphere encapsulates our solar system. The heliosphere is the magnetic bubble inflated from the inside by the high-speed solar wind blowing out from the Sun. Pressure from the solar wind, along with pressure from the surrounding interstellar medium, determines the size and shape of the heliosphere. The supersonic flow of solar wind abruptly slows at the termination shock, the innermost boundary of the solar system. The edge of the solar system is the heliopause. The bow shock pushes ahead through the interstellar medium as the heliosphere plows through the galaxy." [483]

The heliosphere encompasses and interacts with all the magnetic fields of the planets. Scientists at NASA's Voyager mission say, "The Sun's magnetic field extends all the way to the edge of the solar system... Because the Sun spins, its magnetic field becomes twisted and wrinkled... far away from the Sun... the folds bunch up. When a magnetic field gets severely folded... lines of magnetic force criss-cross, and reconnect. The crowded folds... reorganize themselves, sometimes explosively, into foamy magnetic bubbles." [484]

2015 research suggests that the Sun's magnetic field controls the large-scale shape of the heliosphere much more than expected. A new model shows that the magnetic field squeezes the solar wind along the Sun's north-south axis, producing two jets. These jets are then dragged downstream by the flow of the interstellar medium. The model indicates that the heliospheric tail doesn't extend to large distances but is split into two jets. [485]

The heliospheric current sheet is the surface within the solar system where the polarity of the Sun's magnetic field changes from north to south. This field extends throughout the Sun's equatorial plane in the heliosphere. The shape of the current sheet results from the influence of the Sun's rotating magnetic field on the plasma in the interplanetary medium the solar wind. Within the current sheet a small electrical current flows. The thickness of the current sheet is about 10,000 kilometers near the orbit of the Earth. The underlying magnetic field is called the interplanetary magnetic field, and the resulting electric current forms part of the heliospheric current circuit. The heliospheric current sheet is also sometimes called the interplanetary current sheet. [486]

The interplanetary current sheet reaches Earth's ring current where electrons in the solar wind are accelerated into the auroral oval as Earth's magnetosphere comes in contact with interplanetary space and the solar wind. In the outer reaches of the heliosphere where it interacts with interstellar space an auroral oval could exist.

In 2012, the Voyager mission team announced that the Voyager 1 spacecraft had passed out of the heliosphere into interstellar space, traveling further from Earth than any other human made object. In the nearly two years since the announcement there is uncertainty about whether Voyager 1 really crossed the threshold into interstellar space. [487]

In 2014 the scientists predicted that, in the next two years, Voyager 1 will cross the current sheet, the sprawling surface within the heliosphere where the polarity of the Sun's magnetic field changes from plus to minus. The spacecraft will detect a reversal in the magnetic field, proving that it is still within the heliosphere. But, if the magnetic field reversal doesn't happen in the next year or two as expected, that is confirmation that Voyager 1 has already passed into interstellar space. The scientists predict this reversal will most likely happen during 2015, based on observations made by Voyager 1. [488]

Professor Lisa Randall says, "The Oort cloud is a reasonable candidate for the edge of the solar system." [489] Professor Randall goes on to say that although the Oort Cloud is called a cloud it seems to have structure, consisting of a doughnut-shaped inner region and a spherical outer cloud of cometary nuclei in the outer region, which extends much further out. [490] In the Earth chapter up ahead it will be seen that Earth's Van Allen radiation belts are two doughnut-shaped rings that are filled with high-energy electrons. [491]

Professor Lisa Randall says, "The signal that Voyager has reached the edge of the heliosphere and entered outer space would then be its encountering a decrease of charged particles from inside the heliosphere and an increase in those from outside." [492] On September 12, 2013, NASA Scientists decided that a change in magnetic field was actually not required after all when defining the edge of the heliosphere. They decided to go with a less restrictive criterion, which was the increase in electron density by a factor of almost 100, which is expected once outside the heliopause. [493]

Professor Lisa Randall continues, "The solar system encompasses a small fraction of the size of the visible Universe, but it's nonetheless extremely big. By most reasonable measures, it encompasses the Oort cloud, which extends to at least 50,000 times the Earth– Sun distance (1 AU) and very likely twice as far— more than a light-year away. [494] ...The Oort cloud is considered a fairly well-established component of the solar system. [495] ...50,000 AU is about a fifth the distance to the nearest star beyond our solar system. [496] ...One definition for the size of the solar system is characterized by where the magnetic field associated with the solar wind ends— about 15 billion kilometers away, or about 100 AU." [497]

Sources for the Sun and the Heliosphere

[476] 2015, Daily Mail, U.K.
http://www.dailymail.co.uk/sciencetech/article-3031076/Nasa-reveals-plan-send-spacecraft-upper-atmosphere-SUN-time.html#ixzz3WvJkV2HT
[477] 2014, Lund University
http://www.sciencedaily.com/releases/2014/08/140818095204.htm?utm_source=feedburner&utm_medium=email&utm_campaign=Feed%3A+sciencedaily%2Ftop_news%2Ftop_science+%28ScienceDaily%3A+Top+Science+News%29
[478] 2013, University Corporation for Atmospheric Research
https://www.vsp.ucar.edu/Heliophysics/pdf/Lika_sideways_SC.pdf
[479] 2015, Johnson, Brian, Astrology researcher, February 25, 2015
http://thecanadianinstituteforappliedastronomy.yolasite.com/
[480] 2015, Armstrong, Robin, President of the RASA School of Astrology by permission November 2015 http://www.rasa.ws/
[481] 2008, Seymour, Percy *Dark Matters: Unifying Matter, Dark Matter, Dark Energy, and the Universal Grid* (L1017) Kindle Edition.
[482] 2015, Space Weather, http://spaceweather.com/
[483] 2014, Southwest Research Institute and American Geophysical Union
http://www.sciencedaily.com/releases/2014/07/140723114142.htm
[484] 2011, NASA
http://www.nasa.gov/mission_pages/voyager/heliosphere-surprise.html
[485] 2015, University of Maryland
http://www.sciencedaily.com/releases/2015/02/150219115636.htm?utm_source=feedburner&utm_medium=email&utm_campaign=Feed%3A+sciencedaily%2Ftop_news%2Ftop_science+%28ScienceDaily%3A+Top+Science+News%29
[486] 2003, http://en.wikipedia.org/wiki/Heliospheric_current_sheet
[487] 2014, American Geophysical Union
http://www.sciencedaily.com/releases/2014/07/140723114142.htm
[488] 2014, American Geophysical Union
http://www.sciencedaily.com/releases/2014/07/140723114142.htm
[489] 2015, Randall, Lisa, *Dark Matter and the Dinosaurs: The Astounding Interconnectedness of the Universe* Kindle Edition,

Harper Collins. (L 2065)

[490] 2015, Randall, Lisa, *Dark Matter and the Dinosaurs: The Astounding Interconnectedness of the Universe* Kindle Edition. Harper Collins. (L 2031)

[491] University of Colorado at Boulder
http://www.sciencedaily.com/releases/2014/11/141126133829.htm

[492] 2015, Randall, Lisa, *Dark Matter and the Dinosaurs: The Astounding Interconnectedness of the Universe* Kindle Edition. Harper Collins. (L 2090)

[493] 2015, Randall, Lisa, *Dark Matter and the Dinosaurs: The Astounding Interconnectedness of the Universe* Kindle Edition. Harper Collins. (L 2102)

[494] 2015, Randall, Lisa, *Dark Matter and the Dinosaurs: The Astounding Interconnectedness of the Universe* Kindle Edition. Harper Collins. (L 2069)

[495] 2015, Randall, Lisa, *Dark Matter and the Dinosaurs: The Astounding Interconnectedness of the Universe Kindle* Edition. Harper Collins. (L 2041)

[496] 2015, Randall, Lisa, *Dark Matter and the Dinosaurs: The Astounding Interconnectedness of the Universe* Kindle Edition. Harper Collins. (L 2074)

[497] 2015, Randall, Lisa, *Dark Matter and the Dinosaurs: The Astounding Interconnectedness of the Universe* Kindle Edition. Harper Collins. (L 2082)

Inner Planets

Mercury

Mariner 10 discovered Mercury's magnetic field in the 1970s. Mercury's iron core is supposed to have finished cooling long ago and stopped generating magnetism. MESSENGER data from 2011 to 2015 show Mercury's magnetic field appears to be generated by an active dynamo in the planet's core. [498]

MESSENGER orbiting Mercury between 2011 and 2015, made the first in situ observations of Mercury's unique exosphere and ultrathin atmosphere where atoms and molecules are so far apart they are more likely to collide with the surface than with each other. Material is knocked aloft from the surface of Mercury by solar radiation, solar wind bombardment and meteoroid vaporization. MESSENGER found the chemical composition of the exosphere to be hydrogen, helium, sodium, potassium, and calcium. MESSENGER monitored the exosphere as it was stretched out into a comet-like tail as long as 2 million kilometers by the action of the solar wind. This tail, as well as Mercury's magnetic field, was often buffeted by solar activity. [499]

MESSENGER orbited Mercury for four years before crashing into the planet in April 2015. Collected data revealed that Mercury's magnetic field is almost four billion years old. Mercury's magnetic field is similar to Earth's, but much weaker. The motion of liquid iron deep inside the planet's core generates the field. Mercury is the only other planet besides Earth in the inner solar system with such a magnetic field. [500]

Airless Mercury is on average much darker than its closest airless neighbor, Earth's Moon. Comet dust composed of up to 25% carbon expose Mercury to a steady supply of carbon from crumbling comets that often start to break apart as they approach Mercury on their journey to the Sun. [501]

In 2016, scientists led by Dr. Patrick Peplowski, a Research Scientist at the Johns Hopkins University Applied Physics Laboratory, used data from the MESSENGER mission to confirm that a high abundance of carbon is present at Mercury's surface. However, they also have also found that, rather than being delivered by comets, the carbon most likely originated deep below the surface, in the form of a now-disrupted and buried ancient graphite-rich crust, some of which was later brought to the surface by impact processes after most of Mercury's current crust had formed. Deputy Principal Investigator of the MESSENGER mission, Carnegie's Larry Nittler, explained, "The previous proposal of comets delivering carbon to Mercury was based on modelling and simulation. Although we had prior suggestions that carbon may be the darkening agent, we had no direct evidence...The finding of abundant carbon on the surface suggests that we may be seeing remnants of Mercury's original ancient crust mixed into the volcanic rocks and impact ejecta that form the surface we see today." [501a]

The dominant tectonic landforms on Mercury are huge cliffs called lobate scarps. These scarps are signs of global shrinkage, like wrinkles on a raisin. Cooling of Mercury's oversized core has led to a contraction of the planet. [502]

Venus

In 1967, Venera 4 found the magnetic field on Venus to be induced by an interaction between the ionosphere and the solar wind [503] ...rather than by an internal dynamo in the core like the one inside Earth. Venus' small induced magnetosphere provides negligible protection to the atmosphere against cosmic radiation. This radiation may result in cloud-to-cloud lightning discharges. [504]

The existence of lightning on Venus had been controversial since the first suspected bursts were detected by the Soviet Venera probes. In 2006–2007 Venus Express clearly detected whistler-mode waves, the signatures of lightning. Their intermittent appearance indicates a pattern associated with weather activity. [505]

Venus has two vortices (whirlwinds) above its south pole, and two more above its north pole. Astronomers in the Planetary Science Group of the University of the Basque Country have been closely monitoring the complicated movement of the south pole vortices of slow-rotating Venus. The south pole vortex is a huge double whirlwind the size of Europe. In the south polar vortex of Venus, there are two main cloud layers separated by a distance of about 20 kilometers. Scientists announced on March 24, 2013 that they've confirmed the erratic movement of air in the double vortex at Venus' south pole. And, surprisingly they said, each part of the vortex forms a separate tube, which goes its own way. Itziar Garate-Lopez, Head Researcher, said, "The vortices above the poles of Venus appear to be ever-shifting, but permanent." [506]

The weak magnetosphere around Venus means that the solar wind is interacting directly with Venus' outer atmosphere where ions of hydrogen and oxygen are being created by the dissociation of neutral molecules from ultraviolet radiation. The solar wind then supplies energy that gives some of these ions sufficient velocity to escape Venus' gravity field. [507]

Dr. Dean Pesnell, SDO Project Scientist at NASA's Goddard Space Flight Center in Greenbelt, Maryland says that on Venus, "Radiation goes into the atmosphere and is absorbed, creating ions and a layer of the atmosphere called the ionosphere..." [508]

Venus has an ionosphere that is bombarded on the Sun-side of the planet by the solar wind filling the atmosphere with charged particles. The ionosphere is a thin boundary in front of Venus that extends into a long comet-like tail behind. As the solar wind plows into the ionosphere plasma piles up creating a thin magnetosphere around Venus. Pioneer Venus Orbiter noticed a hole in Venus' ionosphere a region where the density just dropped out. Venus Express, launched in 2006, is currently in a 24-hour orbit around the poles of Venus. This orbit places it in much higher altitudes than that of the Pioneer Venus Orbiter and even at those heights the same holes were spotted, thus showing that the holes extended much further into the atmosphere than had been previously known and the holes are more common than realized. Glyn Collinson, a Space Scientist at NASA's Goddard Space Flight Center in Greenbelt, Maryland, says, "We think some of these field lines can sink right through the ionosphere, cutting through it like cheese wire...The ionosphere can conduct electricity, which makes it basically transparent to the field lines." The Venus Express observations suggest that instead of two holes behind Venus, there are in fact two long, fat cylinders of lower density material stretching from the planet's surface to way out into interplanetary space. [509]

On January 29, 2013, European Space Agency scientists reported that the ionosphere of Venus streams outwards in a manner similar to "the ion tail seen streaming from a comet under similar conditions." [510]

On Venus thermal inertia and the transfer of heat by winds in the lower atmosphere mean that the surface temperature does not vary significantly between the night and day sides, despite Venus' extremely slow rotation. Winds at the surface are slow, moving at a few kilometres per hour, but because of the high density of the atmosphere at the surface, they exert a significant amount of force against obstructions, and transport dust and small stones across the surface. [511] Above a dense carbon dioxide layer there are thick clouds, consisting mainly of sulfur dioxide and sulfuric acid droplets. [512] These clouds reflect and scatter about 90% of the sunlight that falls on them back into space, and prevent visual observation of the surface of Venus. The permanent cloud cover means that although Venus is closer than Earth to the Sun, the surface of Venus is not as well lit. Strong 300 kilometer per hour winds at the cloud tops circle the planet. [513] In 2018 the Solar Probe Plus will flyby Venus seven times. [514]

Astrologer John Rutherford says, "With magnetic fields like conditions repel while unlike conditions draw together and attract. Venus rotates retrograde and is astrologically considered the ruler of attractions." [515]

Earth

The Van Allen radiation belts were detected in 1958 by Professor James Van Allen and his team at the University of Iowa and were found to be composed of an inner and outer belt extending up to 25,000 miles above Earth's surface. In 2012 Professor Daniel Baker, director of Colorado University-Boulder's Laboratory for Atmospheric and Space Physics led a team that discovered a third, transient storage ring between the inner and outer Van Allen radiation belts that seems to come and go with the intensity of space weather. [516]

The Van Allen Probes measure particle, electric and magnetic fields, or basically everything in the radiation belt environment, including the electrons, which descend following Earth's magnetic field lines that converge at the poles. Plasma waves buffeting Earth's radiation belts are responsible for scattering charged particles into the atmosphere. These waves are the fallout of electrons from the Van Allen radiation belts. [517] At the region known as the South Atlantic Anomaly the Earth's inner Van Allen radiation belt comes close to the Earth's surface, increasing the flux of energetic particles. [518]

Professor Daniel Baker says, "An invisible shield has been discovered some 7,200 miles above Earth that blocks near-light speed electrons that move around the planet. The barrier to the particle motion was discovered in the Van Allen radiation belts, two doughnut-shaped rings above Earth that are filled with high-energy electrons and protons." The Van Allen radiation belts periodically swell and shrink in response to incoming energy disturbances from the Sun. [521]

Ultrafast electrons are highly energetic particles trapped in Earth's outer radiation belt, which extends from 12,000 kilometers to 64,000 kilometers above the planet's surface. During solar storms the number of electrons, grow at least ten times and they can be dislodged. Ultrafast electrons are energetic enough to cause microscopic lightning strikes. On 7 November 2004, the Sun blasted a solar storm in Earth's direction. It was composed of an interplanetary shock wave followed by a large magnetic cloud. When the shock wave first swept over the ESA-NASA solar watchdog satellite SOHO, the speed of the solar wind (the constant flow of solar particles) suddenly increased from 500 kilometers per second to 700 kilometers per second. Shortly afterwards, the shock wave hit Earth's protective magnetosphere. The impact induced a wave front propagating inside the magnetosphere at more than 1,200 kilometers per second around Earth. The quantity of energetic electrons in the outer radiation belt also increased. Professor Qiugang Zong from Peking University (China) and University of Massachusetts Lowell (USA) says, "Both VLF and ULF waves accelerate electrons in Earth's radiation belts, but with different timescales. The ULF waves are much faster than the VLF, due to their much larger amplitudes." The data show that a two-step process causes the substantial rise of ultrafast electrons. The initial acceleration is due to the strong shock-related magnetic field compression. Immediately after the impact of the interplanetary shock, Earth's magnetic field lines begin wobbling at ultra low frequencies. In turn, these ULF waves effectively accelerate the seed electrons provided by the first step, to become ultrafast electrons. [522]

Earth's Moon

Earth's Moon is mostly made up of mantle and has little to no atmosphere but the magnetic field lines from the Sun go through the Moon's mantle and then hit what is thought to be an iron core. [523] 2014 findings suggest that the interior of the Moon has not yet cooled and hardened, and also that it is still being warmed by the effect of the Earth on the Moon. [524]

According to a dissertation by Dr. Charles Lue at the Swedish Institute of Space Physics and Umea University in Sweden the lunar space environment is much more active than previously assumed. The solar wind is reflected from the surface and local crustal magnetic fields of the Moon which has effects on lunar water levels in the lunar crust. The reflected solar wind ions move in spiraling tracks that can take them from the lunar dayside, where the solar wind strikes first, to the nightside of the Moon. In local areas with strong magnetism, the solar wind flow is restricted on the surface at the same time as adjacent areas receive an increased flow. [525]

Earth/Moon modeling done by University of New Hampshire and NASA scientists suggests that over time periodic storms of solar energetic particles may have significantly altered the properties of the soil in the Moon's coldest craters. The study proposes that high-energy particles from uncommon, large solar storms penetrate the Moon's frigid, polar regions and electrically charge the soil. The charging may create electrostatic breakdown suggesting that permanently shadowed regions may be more active than previously thought. Energetic particles created by solar storms stream through interplanetary space and bombard the Moon. These particles can build up electric charges faster than the soil can dissipate them particularly in the polar cold of permanently shadowed regions as cold as -240 Celsius and known to contain water ice. Dr. Andrew Jordan, a Research Scientist at the UNH Institute for the Study of Earth, Oceans, and Space says, "Breakdown weathering is a process in which electrons, released from the soil grains by strong electric fields, race through the material so quickly that they vaporize little channels. Repeated large solar storms could gradually grow these channels large enough to fragment the grains,

disintegrating the soil into smaller particles of distinct minerals."
[526]

Professor David A. Rothery says, "If there is no atmosphere, solar ultraviolet light can reach the surface, where it may, over time, break chemical bonds. Micrometeorite impacts and...charged particles from the solar wind can also affect surface chemistry, so airless bodies experience a suite of processes, collectively described as space weathering, that slowly alter the composition of the surface. For example, the bonds linking iron to oxygen atoms can be broken, allowing oxygen to escape and leaving submicroscopic grains of pure metal, called nanophase iron." [527]

Mars

The equatorial region of Mars has several very large volcanoes while the two hemispheres of Mars are more different from any other planet in our solar system. The northern hemisphere of Mars is non-volcanic flat lowlands. The southern hemisphere of Mars is highlands with countless volcanoes. [528]

2015 data show that, on Mars, aurora in the upper atmosphere glows blue depending on the activity of the Sun. The strongest colour in the aurora of Mars is deep blue. Green and red also occur, just like on Earth. In fact Mars' upper atmosphere is more like Earth's than previously thought. [529]

NASA's Mars Atmosphere and Volatile Evolution (MAVEN) spacecraft has caught sight of a massive dust cloud, and a glowing aurora not unlike the aurora on Earth. MAVEN also recorded a bright ultraviolet auroral glow in the northern hemisphere of Mars. Like the aurora on Earth, these are produced by solar radiation striking atmospheric particles. Unlike Earth, the aurora on Mars, reach down much lower into the atmosphere. Dr. Arnaud Stiepen of the Maven IUVS team at the University of Colorado says, "The electrons producing aurora must be really energetic." Observations in 2004 by Mars Express were the first time that the aurora phenomenon had been detected. It was found that the auroras on Mars are not like those found anywhere else in the solar system. The aurora, are generated by particle interactions with localized magnetic field emissions, rather than globally generated ones such as the auroras generated here on Earth. [530]

In December 2014, NASA's MAVEN spacecraft detected evidence of widespread auroras in Mars' northern hemisphere. The aurora circled the globe and descended close to the Martian equator. Dr. Nick Schneider, an Associate Professor who leads MAVEN's Imaging Ultraviolet Spectrograph team at the University of Colorado, says, "It really is amazing auroras on Mars appear to be more wide ranging than we ever imagined." Unlike Earth, Mars does not have a global magnetic field that envelops the entire planet. Instead, Mars has umbrella-shaped magnetic fields that sprout out of the ground here and there, but mainly in the southern hemisphere. Dr. Schneider says, "The canopies of the patchwork umbrellas are where we expect to find Martian auroras...MAVEN is seeing aurora outside these umbrellas, so this is something new." Auroras occur, both on Earth and Mars, when energetic particles from space rain down on the upper atmosphere. On Earth, these particles are guided toward the poles by Earth's global magnetic field explaining why auroras are seen most often around the Arctic and Antarctic. On Mars, there is no organized planetary magnetic field to guide the particles north and south—so they can go anywhere. Dr. Schneider says, "The particles seem to precipitate into the atmosphere anywhere they want...Magnetic fields in the solar wind drape across Mars, even into the atmosphere, and the charged particles just follow those field lines down into the atmosphere." [531]

The first comprehensive measurements of the composition of Mars' upper atmosphere and electrically charged ionosphere reveal a new process by which the solar wind can penetrate deep into a planetary atmosphere. The results also offer an unprecedented view of ions as they gain the energy that will lead to their escape from the atmosphere. Professor Bruce Jakosky, a Planetary Scientist and MAVEN Principal Investigator with the Laboratory for Atmospheric and Space Physics at the University of Colorado, Boulder says, "We are beginning to see the links in a chain that begins with solar-driven processes acting on gas in the upper atmosphere and leads to atmospheric loss..." MAVEN's Solar Wind Ion Analyzer has discovered a stream of solar-wind particles that are not deflected but penetrate deep into Mars' upper atmosphere and ionosphere. Interactions in the upper atmosphere appear to transform this stream of ions into a neutral form that can penetrate to surprisingly low altitudes. [532]

MAVEN's Solar Wind Ion Analyzer detected a polar plume of ions escaping from Mars. The energized ions ultimately break free of the gravity of Mars as they move along a plume that extends behind Mars. [533] Dr. Antonio Garcia Munoz, a Research Fellow at ESA's ESTEC says, "Another idea is that plumes are related to an auroral emission, and indeed auroras have been previously observed at these locations, linked to a known region on the surface where there is a large anomaly in the crustal magnetic field..." Plumes have been seen reaching high above the surface of Mars. On two separate occasions in 2012, amateur astronomers reported definite plume-like features developing on the planet. A Professor of Applied Physics, Agustin Sanchez-Lavega of the Universidad del País Vasco in Spain says, "At about 250 kilometers, the division between the atmosphere and outer space is very thin, so the reported plumes are extremely unexpected..." The features developed in less than 10 hours, covering an area of up to 500,000 square kilometers, and remained visible for around 10 days. [534]

Comet C/2013 A1 Siding Spring traveled from the Oort Cloud in the most distant region of our solar system to make a close approach about 87,000 miles from Mars. This is less than half the distance between Earth and our Moon. [535]

The MAVEN research team said that data from observations revealed that debris from comet Siding Spring caused an intense meteor shower on Mars and added a new layer of ions to the ionosphere. The ionosphere is an electrically charged region in the atmosphere that reaches from about 75 miles to several hundred miles above the surface of Mars. Scientists made a direct connection between the input of debris from the meteor shower to the subsequent formation of the transient layer of ions. This is the first time such an event has been observed on any planet, including Earth. [536]

NASA's MAVEN spacecraft observed a spectacular meteor shower from comet Siding Spring. MAVEN did not actually see streaks of light in the atmosphere of Mars as MAVEN was sheltering behind the body of the planet during the comet's flyby. But when MAVEN emerged, it found a glowing layer of ionized magnesium, a constituent of meteor smoke, floating 150 kilometers above the surface of Mars. Dr. Nick Schneider says, "The data are consistent with a few tons of comet dust being deposited in the atmosphere of Mars." [537]

Data from observations carried out by MAVEN et al have revealed that dust from comet Siding Spring impacted Mars and was vaporized high in the atmosphere. This debris resulted in significant temporary changes to the planet's upper atmosphere. The Ultraviolet Spectrograph observed intense ultraviolet emission from magnesium and iron ions high in the atmosphere in the aftermath of the meteor shower. The emission dominated Mars' ultraviolet spectrum for several hours after the encounter and then dissipated over the next two days. [538]

Mars Express observed a huge increase in the density of electrons following Siding Spring's close approach. There was a huge jump in the electron density in the ionosphere a few hours after the comet rendezvous. The increased ionization appears to be the result of fine particles from the comet burning up in the atmosphere. The Shallow Subsurface Radar scientists also detected the enhanced ionosphere and determined that the electron density of the ionosphere on the night side of Mars, where the observations were made, was five to 10 times higher than usual. [539]

Mars Moons

Phobos the larger of two moons is orbiting Mars at 3,700 miles (6,000 kilometers) above the surface of Mars. Phobos is closer to its planet than any other moon in the solar system. [540]

Sources for Inner Planets

[498] 2015, NASA Science http://science.nasa.gov/science-news/science-at-nasa/2015/30apr_messenger/

[499] 2015, NASA Science http://science.nasa.gov/science-news/science-at-nasa/2015/30apr_messenger/

[500] 2015, University of British Columbia http://www.sciencedaily.com/releases/2015/05/150507145200.htm?utm_source=feedburner&utm_medium=email&utm_campaign=Feed%3A+sciencedaily%2Ftop_news%2Ftop_science+%28ScienceDaily%3A+Top+Science+News%29

[501] 2015, Brown University http://www.sciencedaily.com/releases/2015/03/150330122437.htm?utm_source=feedburner&utm_medium=email&utm_campaign=Feed%3A+sciencedaily%2Ftop_news%2Ftop_science+%28ScienceDaily%3A+Top+Science+News%29

[501a] 2016, Carnegie Institution https://www.sciencedaily.com/releases/2016/03/160307112950.htm?utm_source=feedburner&utm_medium=email&utm_campaign=Feed%3A+sciencedaily%2Ftop_news%2Ftop_science+%28ScienceDaily%3A+Top+Science+News%29

[502] 2015, NASA Science http://science.nasa.gov/science-news/science-at-nasa/2015/30apr_messenger/

[503] 1969, 1995, Dolginov, *Nature of the Magnetic Field in the Neighborhood of Venus*, Cosmic Research; Kivelson G. M.; Russell, C. T. *"Introduction to Space Physics"*. Cambridge University Press

[504] 1995, Upadhyay, H. O.; Singh, R. N. *"Cosmic ray Ionization of Lower Venus Atmosphere"* Advances in Space Research 15

[505] 2007, Russell, S. T. Et al. *"Lightning on Venus inferred from whistler-mode waves in the ionosphere"*. Nature 450.

[506] http://earthsky.org/space/surprises-in-venus-south-polar-vortex

[507] 2007, Svedhem, Håkan; et al. *"Venus as a more Earth-like planet"*. Nature 450

[508] 2015, NASA/Goddard Space Flight Center http://www.sciencedaily.com/releases/2015/07/150709180212.htm?utm_source=feedburner&utm_medium=email&utm_campaign=Feed%3A+sciencedaily%2Ftop_news%2Ftop_science+%28ScienceDaily%3A+Top+Science+News%29

[509] 2014, NASA/Goddard Space Flight Center http://www.sciencedaily.com/releases/2014/09/140911180754.htm?utm_source=feedburner&utm_medium=email&utm_campaign=Feed

%3A+sciencedaily%2Ftop_news%2Ftop_science+%28ScienceDaily%3A+Top+Science+News%29

[510] 2013, Staff (January 29, 2013) "When A Planet Behaves Like A Comet". ESA, Kramer, Miriam (January 30, 2013) "Venus Can Have 'Comet-Like' Atmosphere". Space.com.

[511] 1979, Moshkin, B. E.; et al. *"Dust on the surface of Venus"*. Kosmicheskie Issledovaniia (Cosmic Research)

[512] 1981, 2006, Krasnopolsky, V. A.; Parshev, V. A. *"Chemical composition of the atmosphere of Venus"*. Nature 292; *"Chemical composition of Venus atmosphere and clouds: Some unsolved problems"*, Planetary and Space Science 54.

[513] 1990, Rossow, W.B. et al. "Cloud-tracked winds from Pioneer Venus OCPP images"

[514] 2015, Daily Mail, U.K. http://www.dailymail.co.uk/sciencetech/article-3031076/Nasa-reveals-plan-send-spacecraft-upper-atmosphere-SUN-time.html#ixzz3WvJkV2HT

[515] 2015, Astrologer John Rutherford by permission.

[516] 2014, University of Colorado at Boulder http://www.sciencedaily.com/releases/2014/11/141126133829.htm

[517] 2015, Dartmouth College http://www.sciencedaily.com/releases/2015/01/150105125914.htm

[518] http://scitechdaily.com/new-hubblecast-video-explores-south-atlantic-anomaly/

[519] 2007, Akasofu, Syun-Ichi, *Exploring the Secrets of the Aurora* Kindle. (L 592)

[521] 2014, University of Colorado at Boulder http://www.sciencedaily.com/releases/2014/11/141126133829.htm

[522] 2010, European Space Agency http://www.sciencedaily.com/releases/2010/03/100311101659.htm

[523] NASA/Goddard Space Flight Center http://www.sciencedaily.com/releases/2014/09/140911180754.htm?utm_source=feedburner&utm_medium=email&utm_campaign=Feed%3A+sciencedaily%2Ftop_news%2Ftop_science+%28ScienceDaily%3A+Top+Science+News%29

[524] 2014, National Astronomical Observatory of Japan http://www.sciencedaily.com/releases/2014/08/140808110715.htm

[525] 2015, Umea Universitet,
http://www.sciencedaily.com/releases/2015/11/151130084616.htm
[526] 2014, University of New Hampshire
http://www.sciencedaily.com/releases/2014/08/140821102431.htm?
utm_source=feedburner&utm_medium=email&utm_campaign=Feed
%3A+sciencedaily%2Ftop_news%2Ftop_science+%28ScienceDaily
%3A+Top+Science+News%29
[527] 2010, Rothery, David A. Planets: *A Very Short Introduction*
Oxford University Press, Kindle Edition. (L 975)
[528] 2015, ETH Zurich
http://www.sciencedaily.com/releases/2015/01/150128125414.htm?
utm_source=feedburner&utm_medium=email&utm_campaign=Feed
%3A+sciencedaily%2Ftop_news%2Ftop_science+%28ScienceDaily
%3A+Top+Science+News%29
[529] 2015, Aalto University
http://www.sciencedaily.com/releases/2015/05/150527092606.htm?
utm_source=feedburner&utm_medium=email&utm_campaign=Feed
%3A+sciencedaily%2Ftop_news%2Ftop_science+%28ScienceDaily
%3A+Top+Science+News%29
[530] 2015, http://www.cosmosup.com/amazing-nasas-maven-
probe-detected-unexplained-aurora-and-dust-clouds-on-
mars/#sthash.DmQgBf9z.dpuf
[531] 2015, NASA Science http://science.nasa.gov/science-
news/science-at-nasa/2015/11may_aurorasonmars/
[532] 2014, NASA/Goddard Space Flight Center
http://www.sciencedaily.com/releases/2014/12/141215140851.htm?
utm_source=feedburner&utm_medium=email&utm_campaign=Feed
%3A+sciencedaily%2Ftop_news%2Ftop_science+%28ScienceDaily
%3A+Top+Science+News%29
[533] 2014, NASA/Goddard Space Flight Center
http://www.sciencedaily.com/releases/2014/12/141215140851.htm?
utm_source=feedburner&utm_medium=email&utm_campaign=Feed
%3A+sciencedaily%2Ftop_news%2Ftop_science+%28ScienceDaily
%3A+Top+Science+News%29
[534] 2015, European Space Agency
http://www.sciencedaily.com/releases/2015/02/150216200858.htm?
utm_source=feedburner&utm_medium=email&utm_campaign=Feed

%3A+sciencedaily%2Ftop_news%2Ftop_science+%28ScienceDaily%3A+Top+Science+News%29

[535] 2014, NASA http://www.nasa.gov/press/2014/november/mars-spacecraft-reveal-comet-flyby-effects-on-martian-atmosphere/

[536] 2014, University of Colorado at Boulder http://www.sciencedaily.com/releases/2014/11/141107154730.htm?utm_source=feedburner&utm_medium=email&utm_campaign=Feed%3A+sciencedaily%2Ftop_news%2Ftop_science+%28ScienceDaily%3A+Top+Science+News%29

[537] 2015, http://spaceweather.com/

[538] 2014, NASA http://www.nasa.gov/press/2014/november/mars-spacecraft-reveal-comet-flyby-effects-on-martian-atmosphere/

[539] 2014, NASA http://www.nasa.gov/press/2014/november/mars-spacecraft-reveal-comet-flyby-effects-on-martian-atmosphere/

[540] 2015, NASA/Goddard Space Flight Center http://www.sciencedaily.com/releases/2015/11/151110171214.htm?utm_source=feedburner&utm_medium=email&utm_campaign=Feed%3A+sciencedaily%2Ftop_news%2Ftop_science+%28ScienceDaily%3A+Top+Science+News%29

Asteroids and Comets

Asteroids

Professor David A. Rothery says, "The asteroid belt is virtually empty space, and you should not think of it as replete with jostling rocks." [541]

Robert Frost Instructor/Engineer in the Flight Operations Directorate at NASA says, "The average distance between asteroids, in the asteroid belt, is about 1 million kilometers. For comparison, the Moon is 384,000 kilometers from Earth. So if the Earth were an asteroid in the asteroid belt and you looked up to find the closest asteroid, on average, it would be 2.5 times as far away as the Moon is from Earth." [542] Ceres and Vesta are the biggest objects in the asteroid belt.

Researchers at the University of Tennessee, Knoxville studied near-Earth asteroid 1950 DA and discovered that the body, which rotates extremely quickly, is held together by cohesive forces called van der Waals, never detected before on an asteroid. Dr. Ben Rozitis, a Postdoctorial Researcher says, "We found that 1950 DA is rotating faster than the breakup limit for its density...So if just gravity were holding this rubble pile together, as is generally assumed, it would fly apart. Therefore, interparticle cohesive forces must be holding it together." In fact, the rotation is so fast that at its equator, 1950 DA effectively experiences negative gravity. [543]

Comets

The Rosetta spacecraft has been watching the early stages of how a magnetosphere forms around comet 67P Churyumov-Gerasimenko. As the comet gets closer to the Sun it begins to interact with the solar wind. As the comet gets warmer, volatile substances, mainly water, evaporate from the surface and form an atmosphere around the comet. The Sun's ultraviolet radiation and collisions with the solar wind ionizes some of the comet's atmosphere. The newly formed ions are affected by the solar wind electric and magnetic fields and can be accelerated to high speeds. When the comet gets close enough to the Sun, its atmosphere becomes so dense and ionized that it becomes electrically conductive. When this happens, the atmosphere starts to resist the solar wind and a comet's magnetosphere is born. Dr. Hans Nilson, Associate Professor at the Swedish Institute of Space Physics in Kiruna says, "We discovered that the comet atmosphere affects the solar wind more than we thought it would at this early stage. We are also surprised how much structure we see in our data -- the comet atmosphere appears to be very unevenly distributed around the nucleus." [544]

In July, 2015, the European Space Agency reported that the Philae lander from its Rosetta spacecraft in orbit around comet 67P/Churyumov-Gerasimenko detected 16 organic compounds as it descended toward and then bounced across the comet's surface. According to the agency, some of the compounds detected play key roles in the creation of amino acids, nucleobases, and sugars from simpler building-block molecules. [545]

Nicolas Biver, a Researcher at the Paris Observatory, France, lead author of a paper published Oct. 23, 2015 in *Science*

Advances says, "We found that comet Lovejoy was releasing as much alcohol as in at least 500 bottles of wine every second during its peak activity." The team found 21 different organic molecules in gas from the comet, including ethyl alcohol and glycolaldehyde, a simple sugar. Comet Lovejoy (formally cataloged as C/2014 Q2) passed closest to the Sun on January 30, 2015, when it was releasing water at the rate of 20 tons per second. The team observed the atmosphere of the comet around this time when it was brightest and most active. [546]

Dr. Apostolos Christou, a Research Astronomer at the Armagh Observatory in Northern Ireland, Dr. Rosemary Killen at NASA's Goddard Space Flight Center in Greenbelt, Maryland, and Dr. Matthew Burger of Morgan State University in Baltimore say that the planet Mercury is being pelted regularly by bits of dust from an ancient comet. NASA's MESSENGER, the first spacecraft to orbit Mercury, measured how certain species in the exosphere vary with time. Analysis of the data by Dr. Burger and colleagues found a pattern in the variation of the element calcium that repeats from one Mercury year to the next. The peak in calcium emission is seen right after Mercury passes through its perihelion -- the closest point of its orbit to the Sun. Dr. Killen and Dr. Joe Hahn a Research Scientist at the Space Science Institute, based in Austin, Texas proposed that dust from comet Encke impacting Mercury could kick up more calcium from the surface. Dr. Christou found that the dust, rather than shifting away from the comet's orbit, simply spread along it, forming a stream that encounters Mercury exactly when the comet does. Dr. Christou says, "... Encke and its stream from independent sources makes us confident that the cause-and-effect relationship is real," Dr. Killen says, "We already knew that impacts were important in producing exospheres. What we did not know was the relative importance of comet streams over zodiacal dust. Apparently, comet streams can have a huge, but periodic, effect." [547]

Professor Lisa Randall says, "A comet consists of the coma, the nucleus it surrounds, and the tail that streams away...When comets pass into the inner solar system where they are closer to the heat of the Sun do the volatile materials in the comet vaporize so that they stream out, along with some dust, creating an atmosphere around the nucleus called the coma. The coma can be much bigger than the nucleus— thousands or even millions of kilometers across, sometimes even growing to the size of the Sun. Bigger dust particles remain in the coma whereas lighter ones are pushed into the tail by the Sun's radiation and charged particle emissions." [548]

Professor Randall continues, "Meteor showers arise out of the solid debris that comets leave in their wake. The showers occur after a comet has crossed the Earth's orbit so that some of the discarded material lies along the Earth's path. The Earth then passes through the debris on a regular basis, creating periodic meteor showers." [549] ...Comets orbit the Sun with bright ion tails and separate dust tails that generally point in different directions...the dust tail generally follows the comet's path, the ion tail points away from the Sun. The ion tail forms when solar ultraviolet radiation hits the coma, tearing off electrons from some of the atoms. The ionized particles create a magnetic field in what is known as a magnetosphere." [550]

Again Professor Lisa Randall, "In the 1950s the German scientist Ludwig Biermann (and independently another German, Paul Ahnert) made the extraordinary observation that the bright ion tail of a comet always points away from the Sun, Biermann proposed that the Sun emits particles that "push" on the comet tail, making it point this way. In a metaphorical sense, the "solar wind" "blew" the ion tail there...Comet tails can extend up to tens of millions of kilometers." [551]

Also Professor Lisa Randall, "The Stardust spacecraft collected and analyzed dust particles from the coma of the comet Wild 2 in early 2004 and brought this material to Earth for study in 2006. The comet's material did not consist primarily of interstellar medium material as expected for an object formed in the distant Oort cloud, but was instead mostly stuff heated from within the solar system." [552]

Zodiacal light is caused by sunlight reflecting off interplanetary dust particles that orbit the Sun within the inner solar system. [552a] These particles or cosmic dust are considered to be fragments of Jupiter family comets that are found revolving around the Sun in a path between the Sun and Jupiter. These comets have a short revolution period, generally less than 200 years. While cosmic dust can be found throughout our Solar system, zodiacal lights have been mostly observed in the zone around the ecliptic plane. Zodiacal lights are best observed right before sunrise and can sometimes be mistaken as the beginnings of dawn. It is thought that the term false dawn was coined by Persian poet Omar Khayyam in the 12th century. Zodiacal lights have a special significance for practitioners of the Islamic faith where the Prophet Muhammad is known to have used the zodiacal lights to define the times of daily prayers. [552b]

Sources for Asteroids and Comets

[541] 2010, Rothery, David A. *Planets: A Very Short Introduction* Oxford University Press. Kindle Edition. (L 1643)
[542] 2015, Frost, Robert, instructor/engineer in the Flight Operations Directorate at NASA via Quera.com April 8, 2015.

[543] 2014, University of Tennessee
http://www.sciencedaily.com/releases/2014/08/140813132037.htm?
utm_source=feedburner&utm_medium=email&utm_campaign=Feed
%3A+sciencedaily%2Ftop_news%2Ftop_science+%28ScienceDaily
%3A+Top+Science+News%29

[544] 2015, Institutet för rymdfysik - Swedish Institute of Space
Physics (IRF)
http://www.sciencedaily.com/releases/2015/01/150122141802.htm?
utm_source=feedburner&utm_medium=email&utm_campaign=Feed
%3A+sciencedaily%2Ftop_news%2Ftop_science+%28ScienceDaily
%3A+Top+Science+News%29

[545] 2015, NASA/Goddard Space Flight Center
http://www.sciencedaily.com/releases/2015/10/151024092534.htm?
utm_source=feedburner&utm_medium=email&utm_campaign=Feed
%3A+sciencedaily%2Ftop_news%2Ftop_science+%28ScienceDaily
%3A+Top+Science+News%29

[546] 2015, NASA/Goddard Space Flight Center
http://www.sciencedaily.com/releases/2015/10/151024092534.htm?
utm_source=feedburner&utm_medium=email&utm_campaign=Feed
%3A+sciencedaily%2Ftop_news%2Ftop_science+%28ScienceDaily
%3A+Top+Science+News%29

[547] 2015, NASA/Goddard Space Flight Center
http://www.sciencedaily.com/releases/2015/11/151110171351.htm?
utm_source=feedburner&utm_medium=email&utm_campaign=Feed
%3A+sciencedaily%2Ftop_news%2Ftop_science+%28ScienceDaily
%3A+Top+Science+News%29

[548] 2015, Randall, Lisa, *Dark Matter and the Dinosaurs: The
Astounding Interconnectedness of the Univer*se Kindle Edition.
Harper Collins. (L 1786)

[549] 2015, Randall, Lisa, *Dark Matter and the Dinosaurs: The
Astounding Interconnectedness of the Universe* Kindle Edition.
Harper Collins. (L 1791)

[550] 2015, Randall, Lisa, *Dark Matter and the Dinosaurs: The
Astounding Interconnectedness of the Universe* Kindle Edition
Harper Collins. (L 1798)

[551] 2015, Randall, Lisa, *Dark Matter and the Dinosaurs: The
Astounding Interconnectedness of the Universe* Kindle Edition.
Harper Collins. (L 1805)

[552] 2015, Randall, Lisa, *Dark Matter and the Dinosaurs: The Astounding Interconnectedness of the Universe* Kindle Edition. Harper Collins. (L 1848)
[552a] 2015, http://earthsky.org/?p=57881
[552b] 2015, http://www.timeanddate.com/astronomy/zodiacal-lights.html

Outer Planets

Jupiter

Jupiter's thermosphere demonstrates such phenomena as air glow and polar aurora. [553] The energetic particles coming from Jupiter's magnetosphere create bright auroral ovals, which encircle the poles of Jupiter. On Earth aurora appear only during magnetic storms. The aurora on Jupiter, are permanent features of Jupiter's atmosphere. [554]

The interior of Jupiter is fluid and lacks any solid surface. The atmosphere of Jupiter lacks a clear lower boundary and gradually transitions into the liquid interior of the planet. From the highest to the lowest, the atmospheric layers are the exosphere, thermosphere, stratosphere, and the troposphere. [555] The vertical temperature variations in the atmosphere of Jupiter are similar to those of the atmosphere of Earth. The temperature of the troposphere decreases with height until it reaches a minimum at the tropopause. [556] The density also gradually decreases until it smoothly transitions into the interplanetary medium. [557]

Jupiter has powerful storms, often accompanied by lightning strikes. The storms are a result of moist convection in the atmosphere connected to the evaporation and condensation of water. There are sites of strong upward motion of the air, which leads to the formation of bright and dense clouds. The storms form mainly in belt regions. The lightning strikes on Jupiter are hundreds of times more powerful than those seen on Earth. But the amount of lightning activity is comparable to Earth. [558]

The upper ammonia clouds visible at Jupiter's surface are organized in a dozen zonal bands parallel to the equator and are bounded by powerful zonal atmospheric wind flows. [559] These belts and zones that divide Jupiter's atmosphere each have their own unique characteristics. [560] The origin of Jupiter's banded structure is not completely clear, though it may be similar to that driving the Earth's Hadley cells. The simplest interpretation is that zones are sites of atmospheric upwelling, whereas belts are manifestations of downwelling. [561]

The water clouds form the densest layer of clouds and have the strongest influence on the dynamics of the atmosphere. [562] Computer simulations show that Jupiter ejected more water-rich material than it scattered inward. [563]

Jupiter Moons

2015 observations of Jupiter's extreme ultraviolet emissions show that bright explosions of Jupiter's aurora likely get started by the planet-moon interaction, not by solar activity. [564] Hubble has tracked Jupiter's moon Ganymede and has shown how the magnetic field draws in and excites space particles, generating a glow of ultraviolet light around the north and south poles. [565]

A 2015 finding confirms that Jupiter's moon Ganymede has its own global salty ocean below a frozen rocky crust. The ocean is electrically conductive that is evidenced by aurora. [566]

Io, the innermost of Jupiter's four large Galilean moons, is about 2,300 miles across. Io is the only known place in the solar system with volcanoes that are almost continuously erupting extremely hot lava like that on Earth. Because of Io's low gravity, large eruptions produce an umbrella of debris that rises high into space. Recent eruptions match past events that spewed tens of cubic miles of lava over hundreds of square miles in a short period of time. Dr. Ashley Davies, a Volcanologist with NASA's Jet Propulsion Laboratory in Pasadena, California, says, "These new events are in a relatively rare class of eruptions on Io because of their size and astonishingly high thermal emission...The amount of energy being emitted by these eruptions implies lava fountains gushing out of fissures at a very large volume per second, forming lava flows that quickly spread over the surface of Io." [567]

Saturn

Dr. Randy Russell at UCAR Center for Science Information says, "Saturn's magnetosphere, like Earth's, produces aurora." [568]

Saturn has an intrinsic magnetic field. Its strength at the equator is approximately one twentieth of that of the field around Jupiter and slightly weaker than Earth's magnetic field. [569] Matthew McDermott at Thinkquest internet Challenge says, "Most probably, the magnetic field on Saturn is generated similarly to that of Jupiter – by currents in the liquid metallic-hydrogen layer called a metallic-hydrogen dynamo." [570]

Saturn has a pale yellow hue due to ammonia crystals in its upper atmosphere. Electrical current within the metallic hydrogen layer is thought to give rise to Saturn's planetary magnetic field, which is weaker than Earth's, but has a magnetic field 580 times that of Earth due to Saturn's larger size. Saturn's magnetic field strength is around one-twentieth the strength of Jupiter's. [571] Wind speeds on Saturn can reach 1,800 kilometers per hour, faster than on Jupiter, but not as fast as those on Neptune. [572]

Standard planetary models suggest that the interior of Saturn is similar to that of Jupiter, having a small rocky core surrounded by hydrogen and helium with trace amounts of various volatiles. [573] There is potentially conflicting information about Jupiter when one source says, "the interior of Jupiter is fluid and lacks any solid surface" and another says, "similar to that of Jupiter, having a small rocky core". But perhaps not, it seems that Jupiter can have a small core but also have no surface.

Ultraviolet radiation from the Sun causes methane photolysis in Saturn's upper atmosphere, leading to a series of hydrocarbon chemical reactions with the resulting products being carried downward by eddies and diffusion. This photochemical cycle is modulated by Saturn's annual seasonal cycle. [574]

Saturn is a gas giant because it is predominantly composed of hydrogen and helium. It lacks a definite surface, though it may have a solid core. [575] Saturn's interior is probably composed of a core of iron-nickel and rock. This core is surrounded by a deep layer of metallic hydrogen, an intermediate layer of liquid hydrogen and liquid helium, and a gaseous outer layer. [576] The particles that make up Saturn's rings travel between 20,000 to 53,000 miles per hour. The rings vary in thickness, averaging about 30 feet thick. [577]

Since early 2005, scientists have been tracking lightning on Saturn. The power of the lightning is approximately 1,000 times that of lightning on Earth. [578]

The moon Titan orbits within the outer part of Saturn's magnetosphere and contributes plasma from the ionized particles in Titan's outer atmosphere. [579]

Saturn Moons

Kirk Munsell, Science Writer at the NASA Jet Propultion Laboratory says, "Titan, Saturn's largest moon is larger than the planet Mercury and is the only moon in the solar system to have a substantial atmosphere." [580]

Professor David A. Rothery says, "Titan's surface geological processes are similar to many of the processes that occur on Earth [581] ...Titan is 5,150 kilometres in diameter with a dense atmosphere that is 97% nitrogen made opaque by methane and its photochemical derivatives, that turn the stratosphere into an opaque smog. Titan has a crust and mantle made of ice that is (mostly water-ice) occupying the outer one-third of Titan's radius and overlying a rocky core. There could be an iron inner core [582] ...Rainfall on Titan must consist of droplets of methane that, like rainfall on Earth, infiltrates the ground and feeds springs that supply streams and rivers...Mars had rainfall, rivers, and lakes long ago, but Titan is the only other place where they occur today." [583]

Climate simulations indicate that Titan's near-surface winds blow toward the west but surface dunes, reaching a hundred yards high and many miles long, point to the east. The attitude of Titan's sand dunes results from rare methane storms that produce eastward gusts much stronger than the usual westward surface winds. University of Washington Astronomer Dr. Benjamin Charnay said that their model suggests that these methane storms "produce strong downdrafts, flowing eastward when they reach the surface...thus rearranging the dunes...These fast eastward gusts dominate the sand transport, and thus dunes propagate eastward..." The storm winds reach up to 22 miles per hour, about 10 times faster than Titan's near-surface winds. And though the storms happen only when Titan is in equinox and its days and nights are of equal length, about every 14.75 years, they are of sufficient power to realign Titan's dunes. Titan was last in equinox in August 2009. According to Cassini's observations, Titan's atmosphere is in super-rotation above about 5 miles, meaning that it rotates a lot faster than the surface itself. [584]

Scientists have found that the interactions between Titan's atmosphere, and the solar magnetic field and radiation, create a wind of hydrocarbons and nitriles that blow away from Titan's polar regions into interplanetary space. This is very similar to the wind observed coming from the Earth's polar regions. Professor Andrew Coates of UCL Mullard Space Science Laboratory says that data proved that the top of Titan's atmosphere is losing about seven tonnes of hydrocarbons and nitriles every day. Professor Coates further explains that this atmospheric loss is driven by a polar wind powered by an interaction between sunlight, the solar magnetic field and the molecules present in the upper atmosphere. Titan has no magnetic field of its own, but is surrounded by Saturn's rapidly rotating magnetic field, which forms a comet-like tail around Titan. Negatively-charged photoelectrons, spread throughout Titan's ionosphere and the tail, sets up an electrical field. The electrical field, in turn, is strong enough to pull the positively charged hydrocarbon and nitrile particles from the atmosphere throughout the sunlit portion of the atmosphere, setting up the widespread polar wind. This phenomenon has only been observed on Earth before, in the polar regions where Earth's magnetic field is open. As Titan lacks its own magnetic field the same thing can occur over wider regions, not just near the poles. A similarly widespread polar wind is strongly suspected to exist both on Mars and Venus. It gives further evidence of how Titan, despite its location in orbit around a gas giant in the outer solar system, is one of the most Earth-like objects ever studied. [585]

Scientists have detected a monstrous new cloud of frozen compounds in Titan's low- to mid-stratosphere peaking at an altitude of about 124 miles. Circulation in the atmosphere transports gases from the pole in the warm hemisphere to the pole in the cold hemisphere. At the cold pole, the warm air sinks. Scientists determined that temperatures at the south pole must get down to at least -238 Fahrenheit (-150 Celsius). The new cloud was found in the lower stratosphere, where temperatures are even colder. [586]

NASA's Cassini probe in 2015 discovered that Saturn's moon Enceladus has a shallow global ocean beneath a 30 mile thick icy crust. At Enceladus's south polar region it has about 100 saltwater geysers erupting from the surface that contribute nanometer-size silicate particles to one of Saturn's rings. [587]

One of the mysteries this gives us clues to answering is how Saturn's magnetic bubble, known as its magnetosphere, gets rid of gas from Saturn's tiny icy moon Enceladus. Through jets at its south pole, this tiny 500 kilometers sized moon ejects around 100 kilograms of water into space every second. Dr. Chris Arridge a Space Physicist says, "Water from the Enceladus plume is trapped in Saturn's magnetosphere. We know it can't just stay there forever and until now we have not been able to work out how it has been ejected from the magnetosphere." [588]

Uranus

In 2011 astronomers caught the first views of auroras on Uranus. The Uranus aurora photos were captured by the Hubble Space Telescope, marking the first time Uranus' aurora has been seen by an observatory near Earth. [589] The ultraviolet images were taken at the time of heightened solar activity in November 2011 that successively buffeted the Earth, Jupiter, and Uranus with a greatly increased flow of charged particles from the Sun. Because Uranus' magnetic field is inclined 59 degrees to its spin axis, the auroral spots appear far from the planet's north and south poles. [590]

Astrologer John Rutherford says, "With magnetic fields like conditions repel while unlike conditions draw together and attract. Uranus is tipped over, at almost a right angle. When magnetic fields are at right angles to each other, the field lines of one go through those of the other as if they didn't exist. Astrologically Uranus rules freedom and independence. And, Uranus rules politics, a loose and fragile collection of like-minded people." [591]

The intensity of the magnetic field at Uranus' surface is roughly comparable to that of Earth's. The source of Uranus' magnetic field is unknown and the electrically conductive, super-pressurized ocean of water and ammonia once thought to lie between the core and the atmosphere now appears to be nonexistent. Voyager 2 found radiation belts at Uranus with intensity similar to those at Saturn, although they differ in composition, these radiation belts appear to be dominated by hydrogen ions. [592]

The gas-giant Uranus is also considered by NASA to be an ice giant. The winds are as high as 800 kilometers an hour and there are massive thunderstorms of an extreme quality. Uranus has an axial tilt of 97.77 degrees that gives seasonal changes completely unlike those of the other major planets. Near the time of Uranian solstices, one pole faces the Sun continuously and the other one faces away. Only a narrow strip around the equator experiences a rapid day/night cycle. At the other side of Uranus' orbit the orientation of the poles towards the Sun is reversed. Each pole gets around 42 years of continuous sunlight, followed by 42 years of darkness. [593]

Lawrence Sromovsky and Patrick Fry et al from the University of Wisconsin-Madison say, "The climate of Uranus is heavily influenced by both its lack of internal heat, which limits atmospheric activity, and by its extreme axial tilt, which induces intense seasonal variation. Uranus' atmosphere is remarkably bland in comparison to the other gas giants which it otherwise closely resembles [594] ...Zonal winds blow in the upper troposphere of Uranus...At the equator winds are retrograde blowing in the reverse direction to the planetary rotation [595] ...Closer to the poles, the winds shift to a prograde direction, flowing with Uranus' rotation. Wind speeds continue to increase reaching maxima at about 60° latitude before falling to zero at the poles." [596]

In 2006 Uranus displayed elevated levels of brightness, which suggests that the north pole was not always so dim. [597] Uranus is an ellipse rotating about its minor axis which causes its visible area to become larger when viewed from the poles. This explains in part its brighter appearance at solstices. [598] Collected information implies that the visible pole brightens some time before the solstice and darkens after the equinox. [599] Uranus is also known to exhibit strong meridional variations. [600] The south polar region of Uranus is much brighter than the equatorial bands. [601] ...In addition, both poles demonstrate elevated brightness in the microwave part of the spectrum. [602] The lowest temperature recorded in Uranus' tropopause is -224 Celsius, making Uranus the coldest planet in the solar system, colder than Neptune. [603]

Lawrence Sromovsky and Patrick Fry et al say, "Uranus' internal heat appears markedly lower than that of the other giant planets; in astronomical terms, it has a low internal thermal flux [604] ...Why Uranus' heat flux is so low is still not understood. Neptune, which is Uranus' near twin in size and composition, radiates 2.61 times as much energy into space as it receives from the Sun. [605] Uranus, by contrast, radiates hardly any excess heat at all. [606]

Neptune

Voyager detected auroras, similar to aurora on Earth, in Neptune's atmosphere. The auroras on Earth occur when energetic particles strike the atmosphere as they spiral down the magnetic lines. Due to Neptune's complex magnetic field the auroras are an extremely complicated processes that occurs over wide regions of Neptune not just near the magnetic poles. The auroral power on Neptune appears weak, estimated to be at about 50 million watts, compared to 100 billion watts on Earth. [607]

Neptune has a magnetosphere, with a magnetic field strongly tilted relative to its rotational axis. This field may be generated by convective fluid motions in a thin spherical shell of electrically conducting liquids [608] resulting in a dynamo action. [609]

Neptune's bow shock, where the magnetosphere begins to slow the solar wind, occurs at a distance of about 35 times the radius of the planet. The magnetopause, where the pressure of the magnetosphere counterbalances the solar wind, lies at a distance of about 25 times the radius of Neptune. The tail of the magnetosphere extends out to at least 72 times the radius of Neptune, and possibly much further. [610]

Gas giant Neptune's atmosphere has active and visible weather patterns. When Voyager 2 flew by in 1989 Neptune's southern hemisphere had a great dark spot comparable to the great red spot on Jupiter. These weather patterns are driven by the strongest sustained winds of any planet in the solar system, with recorded wind speeds as high as 2,100 kilometres per hour. Neptune's outer atmosphere is one of the coldest places in the solar system with temperatures at its cloud tops approaching -218 Celsius. [611]

Neptune's orbit has a profound impact on the Kuiper belt a ring of small icy worlds, similar to the asteroid belt but far larger, extending from Neptune's orbit at 30 AU* out to about 55 AU* from the Sun. [612] *An AU or Astronomical Unit is about the distance from Earth to the Sun.

Neptune Moons

Professor David A. Rothery says, "Triton, Neptune's largest moon has geysers that erupt through the polar cap, lofting dark particles to a height of about 8 kilometres. There are a few high-altitude clouds made of nitrogen crystals, analogous to cirrus clouds in our own atmosphere." [613]

Pluto

In July 2015 the New Horizons Atmospheres team observed Pluto's atmosphere as far as 1,000 miles above the surface, demonstrating that Pluto's nitrogen-rich atmosphere is quite extended. The New Horizons Particles and Plasma team discovered a region of cold, dense ionized gas tens of thousands of miles beyond Pluto suggesting that the planet's atmosphere is being stripped away by the solar wind and lost to interplanetary space. [614]

New Horizons in July 2015 discovered a region of cold, dense ionized gas tens of thousands of miles beyond Pluto where Pluto's atmosphere has been stripped away by the solar wind and lost to space. The Solar Wind Around Pluto (SWAP) instrument observed a cavity in the solar wind's outflow of electrically charged particles between 48,000 miles and 68,000 miles downstream of Pluto. SWAP data revealed this cavity to be populated with nitrogen ions forming a plasma tail of undetermined structure and length extending behind the planet. Similar plasma tails are observed at planets like Venus and Mars. In the case of Pluto's predominantly nitrogen atmosphere, escaping molecules are ionized by solar ultraviolet light that is picked up by the solar wind, and carried past Pluto to form the plasma tail discovered by New Horizons. Prior to closest approach, nitrogen ions were detected far upstream of Pluto by the Pluto Energetic Particle Spectrometer Science Investigation instrument, providing a foretaste of Pluto's escaping atmosphere. [615]

In the 2015 rapidly updating picture, Pluto is the leading representative of the Kuiper Belt. Unlike any of the other objects out there, Pluto has a wispy but enormous atmosphere composed of nitrogen, methane and other gases that evaporate off its surface when it nears the Sun. New Horizons Co-Investigator Dr. Michael Summers estimates the volume of the atmosphere is 350 times the volume of Pluto itself. And the vague markings visible from Earth show that Pluto has extreme contrasts of light and dark, indicating highly varied terrain. [616]

New Horizons scientists are finding how Pluto and its moons interact with the solar wind, the constant stream of particles and plasma that flows from the Sun and is still travelling at 900,000 miles per hour at Pluto. Pluto's out flowing atmosphere provides a source of neutral atoms that can exchange electrons with the solar wind's positively charged atoms of oxygen, carbon, and nitrogen. Team members searched for X-ray emissions near Pluto to help determine the rate at which Pluto's atmosphere is being lost to space, in much the same way X-ray emissions are used to characterize the outflow of material from comets. [617]

Pluto Moons

Pluto has an intricate family of five known satellites, including the largest Charon, which is nearly half Pluto's diameter. Dr. Alan Stern Principal Investigator for New Horizons says, "On its own, Charon is one of the largest objects in the Kuiper Belt..." [618] Charon orbits quite close to Pluto, about 12,000 miles away. [619]

Dr. John Grunsfeld, Associate Administrator of NASA's Science Mission Directorate in Washington, D.C. says, "Hubble has provided a new view of Pluto and its moons revealing a cosmic dance with a chaotic rhythm...When the New Horizons spacecraft flies through the Pluto system in July 2015 we'll get a chance to see what these moons look like up close and personal...the moons are embedded inside a dynamically shifting gravitational field caused by the system's two central bodies, Pluto and Charon, whirling about each other. The variable gravitational field induces torques that send the smaller moons tumbling in unpredictable ways. This torque is strengthened by the fact the moons are football shaped rather than spherical." [620]

Sources for Outer Planets

[553] 2004, Yelle, R.V.; Miller, S. *"Jupiter's Thermosphere and Ionosphere"*
[554] 2000, Bhardwaj, Anil et al. *Auroral Emissions of the Giant Planets*
[555] 1999, Guillot, T. *"A comparison of the interiors of Jupiter and Saturn"*. https://en.wikipedia.org/wiki/Atmosphere_of_Jupiter
[556] 1969, Ingersoll, A.P.; Cuzzi, J.N. *"Dynamics of Jupiter's cloud bands"*, Journal of the Atmosphereic Sciences
[557] 2005, Miller, Steve; et al. *"Giant Planet Ionospheres and Thermospheres: The Importance of Ion-Neutral Coupling"*.
[558] 2005, Vasavada, A.R.; Showman, A. *"Jovian atmospheric dynamics: An update after Galileo and Cassini"*.
[559] 2014, https://en.wikipedia.org/wiki/Atmosphere_of_Jupiter
[560] 2014, https://en.wikipedia.org/wiki/Atmosphere_of_Jupiter
[561]2004, Ingersoll, A.P.; Dowling, T.E.; Gierasch, P.J. et al. *"Dynamics of Jupiter's Atmosphere"*
[562] 2014, https://en.wikipedia.org/wiki/Atmosphere_of_Jupiter
[563] 2007, Catling, David C. *Astrobiology: A Very Short Introduction* Oxford University Press Kindle Edition. (L 547)
[564] 2015, American Geophysical Union http://www.sciencedaily.com/releases/2015/03/150325110823.htm?utm_source=feedburner&utm_medium=email&utm_campaign=Feed%3A+sciencedaily%2Ftop_news%2Ftop_science+%28ScienceDaily%3A+Top+Science+News%29
[565] 2015, Amos, Jonathan, BBC Science Correspondent http://www.bbc.com/news/science-environment-31855395

[566] 2016 January/February Discover Magazine
http://discovermagazine.com/2016/janfeb/44-saturns-watery-moon
[567] 2014, NASA/Jet Propulsion Laboratory
http://www.sciencedaily.com/releases/2014/08/140804141013.htm?
utm_source=feedburner&utm_medium=email&utm_campaign=Feed
%3A+sciencedaily%2Ftop_news%2Ftop_science+%28ScienceDaily
%3A+Top+Science+News%29
[568] 2003, Russell, Randy (3 June 2003) Windows to the Universe
http://staff.ucar.edu/browse/orgs/SciEd
[569] 1997, Russell, C. T. et al. UCLA – IGPP Space Physics
Center
[570] 2000, McDermott, Matthew, Thinkquest Internet Challenge.
[571] 1997, Russell, C. T. et al. UCLA – IGPP Space Physics
Center.
[572] 2004, "The Planets ('Giants')", Science Channel. 8 June 2004.
[573] 2009, Guillot, Tristan et al. (2009) "Saturn's Exploration
Beyond Cassini-Huygens" Dougherty, Michele K.; et al. Saturn from
Cassini-Huygens. Springer Science+Business Media B.V. p. 745
[574] 2008, Guerlet, S. Et al.
"Ethane, acetylene and propane distribution in Saturn's stratosphere
from Cassini/CIRS limb observations"
[575] 2011, Melosh, H. Jay, Planetary Surface Processes.
Cambridge Planetary Science 13. Cambridge University Press. p. 5.
[576] 2004, Brainerd, Jerome James (27 October 2004) "Giant
Gaseous Planets". The Astrophysics Spectator. Archived.
[577] 2015, Frost, Robert, engineer/instructor NASA, via
www.quera.com February 22, 2015
[578] 2007, "Astronomers Find Giant Lightning Storm At Saturn".
ScienceDaily
[579] 1997, Russell, C. T. et al. "Saturn: Magnetic Field and
Magnetosphere". UCLA – IGPP Space Physics Center.
[580] 2005, Munsell, Kirk, "The Story of Saturn". NASA Jet
Propulsion Laboratory; California Institute of Technology
[581] 2010, Rothery, David A. Planets: A Very Short Introduction
Oxford University Press. Kindle Edition. (L 1532)
[582] 2010, Rothery, David A. Planets: A Very Short Introduction
Oxford University Press. Kindle Edition. (L 1521)

[583] 2010, Rothery, David A. *Planets: A Very Short Introduction* Oxford University Press. Kindle Edition. (L 1549)

[584] 2015, University of Washington
http://www.sciencedaily.com/releases/2015/04/150413183738.htm?utm_source=feedburner&utm_medium=email&utm_campaign=Feed%3A+sciencedaily%2Ftop_news%2Ftop_science+%28ScienceDaily%3A+Top+Science+News%29

[585] 2016, University College London
http://www.sciencedaily.com/releases/2015/06/150618103830.htm?utm_source=feedburner&utm_medium=email&utm_campaign=Feed%3A+sciencedaily%2Ftop_news%2Ftop_science+%28ScienceDaily%3A+Top+Science+News%29

[586] 2015, NASA/Goddard Space Flight Center
http://www.sciencedaily.com/releases/2015/11/151112123531.htm?utm_source=feedburner&utm_medium=email&utm_campaign=Feed%3A+sciencedaily%2Ftop_news%2Ftop_science+%28ScienceDaily%3A+Top+Science+News%29

[587] 2016, January/February Discover Magazine
http://discovermagazine.com/2016/janfeb/44-saturns-watery-moon

[588] 2015, Lancaster University
http://www.sciencedaily.com/releases/2015/12/151201094239.htm?utm_source=feedburner&utm_medium=email&utm_campaign=Feed%3A+sciencedaily%2Ftop_news%2Ftop_science+%28ScienceDaily%3A+Top+Science+News%29

[589] 2012, http://www.space.com/15270-auroras-uranus-hubble-telescope-photos.html

[590] 2012, http://www.nasa.gov/mission_pages/hubble/science/uranus-aurora.html

[591] 2015, Astrologer John Rutherford by permission.

[592] 2014, http://voyager.jpl.nasa.gov/science/uranus_magnetosphere.html

[593] 2014, https://en.wikipedia.org/wiki/Climate_of_Uranus

[594] 2010, Sromovsky, L. A. Et al. *"Dynamics of cloud features on Uranus"* Pierrehumbert, Raymond T. "Principles of Planetary Climate". Cambridge University Press, p. 20

[595] 2005, Sromovsky, L. A. Et al. *"Dynamics of cloud features on Uranus"* Hammel, H. B. Et al. "Uranus in 2003: Zonal winds, banded structure, and discrete features"

[596] 2005, Sromovsky, L. A.; Fry, P. M. *"Dynamics of cloud features on Uranus".*

[597] 2006, Lockwood, G. W.; Jerzykiewicz, M. A. A. *"Photometric variability of Uranus and Neptune",* 1950–2004".

[598] 2006, Lockwood, G. W.; Jerzykiewicz, M. A. A. *"Photometric variability of Uranus and Neptune",* 1950–2004".

[599] 2007, Hammel, H. B.; Lockwood, G. W. *"Long-term atmospheric variability on Uranus and Neptune".* Icarus 186

[600] 2001, Karkoschka, Erich *"Uranus"*

[601] 1986, Smith, B. A.; et al. *"Voyager 2 in the Uranian System: Imaging Science Results".* Science 233

[602] 2003, Hofstadter, M. D.; Butler, B. J. *"Seasonal change in the deep atmosphere of Uranus".* Icarus 165

[603] 1990, 1993, Pearl, J. C. Et al. *"The albedo, effective temperature, and energy balance of Uranus, as determined from Voyager IRIS data",* Icarus 84.
Lunine, Jonathan I. *"The Atmospheres of Uranus and Neptune".* Annual Review of Astronomy and Astrophysics 31:

[604] 2005, 1986, Sromovsky, L. A.; Fry, P. M. *"Dynamics of cloud features on Uranus"* Hanel, R. Et al. *"Infrared Observations of the Uranian System".* Science 233

[605] 2005, Sromovsky, L. A.; Fry, P. M. *"Dynamics of cloud features on Uranus".*

[606] 2014, https://en.wikipedia.org/wiki/Climate_of_Uranus

[607] 2014, http://voyager.jpl.nasa.gov/science/neptune_magnetic.html

[608] 2006, Elkins-Tanton, Linda T. *Uranus, Neptune, Pluto, and the Outer Solar System.* New York: Chelsea House. pp. 79–83

[609] 2004, Stanley, Sabine; Bloxham, Jeremy, Nature 428 *"Convective-region geometry as the cause of Uranus' and Neptune's unusual magnetic fields".*

[610] 1989, Ness, N. F. et al. Science 246 *"Magnetic Fields at Neptune".*

[611] 2015, Kramer, Manfred, Feb 11, 2015. www.Quora.com

[612] 1997, Stern, S. Alan; Colwell, Joshua E.
The Astrophysical Journal, Southwest Research Institute, 490
"Collisional Erosion in the Primordial Edgeworth-Kuiper Belt and the Generation of the 30–50 AU Kuiper Gap".
[613] 2010, Rothery, David A. Planets: *A Very Short Introduction*
Oxford University Press, Kindle Edition. (L 1587)
[614] 2015, NASA
http://www.sciencedaily.com/releases/2015/07/150717174649.htm?utm_source=feedburner&utm_medium=email&utm_campaign=Feed%3A+sciencedaily%2Ftop_news%2Ftop_science+%28ScienceDaily%3A+Top+Science+News%29
[615] 2015, NASA Science
http://solarsystem.nasa.gov/multimedia/display.cfm?IM_ID=20271
[616] 2015, Discover Magazine
http://discovermagazine.com/2015/july-aug/1-pluto-focus
[617] 2015, AstronomyNow
https://astronomynow.com/2015/12/20/new-findings-on-pluto-and-its-moons-from-new-horizons/
[618] 2015, Discover Magazine
 http://discovermagazine.com/2015/july-aug/1-pluto-focus
[619] 2015, Kramer, Manfred, Feb 11, 2015, www.Quora.com
[620] 2015, Space Telescope Science Institute
http://www.sciencedaily.com/releases/2015/06/150603130447.htm?utm_source=feedburner&utm_medium=email&utm_campaign=Feed%3A+sciencedaily%2Ftop_news%2Ftop_science+%28ScienceDaily%3A+Top+Science+News%29

Life

Professor David C. Catling says, "...how life arose is unknown [621] ...It may have originated on Earth or it was carried here by space dust or meteorites [622] ...Many of the problems in defining life boil down to the fact that we have only one example— life on Earth. All Earth-based organisms use nucleic acids for hereditary information, proteins to control biochemical reaction rates, and identical phosphorus-containing molecules to store energy [623] ...But chemical systems need several other features before they can be considered alive. The main ones are a metabolism, enclosure by a semi-permeable membrane, and reproduction that incorporates heredity with a genome." [624]

Astrologer Isabel Hickey said, "The birth-chart indicates certain physical, mental, emotional, and spiritual tendencies with which the person is endowed at birth. [624a] ...The blueprint we call the horoscope, or the birthchart, plots the energies that flow in your magnetic field. At the moment of birth you took into your body, with the first breath, the vibrations manifested on that day and time at that particular spot on earth. This basic pattern goes with you throughout life." [624b]

Professor Kenneth Nealson at the University of Southern California, Los Angeles says that we know that life, when you boil it right down, is a flow of electrons, "You eat sugars that have excess electrons, and you breathe in oxygen that willingly takes them." Our cells break down the sugars, and the electrons flow through the cells in a complex set of chemical reactions until they are passed on to electron-hungry oxygen. In the process, cells make ATP, a molecule that acts as an energy storage unit for almost all living things. Moving electrons around is a key part of making ATP. Professor

Nealson continues, "Life's very clever...It figures out how to suck electrons out of everything we eat and keep them under control." In most living things, the body packages the electrons up into molecules that can safely carry them through the cells until they are dumped on to oxygen. Professor Nealson again, "That's the way we make all our energy and it's the same for every organism on this planet...Electrons must flow in order for energy to be gained. This is why when someone suffocates another person they are dead within minutes. You have stopped the supply of oxygen, so the electrons can no longer flow." [625]

Astrologer John Rutherford says, "All life is about the flow of charged particles. I would say life is the continuous flow of ionic charges, but since electrons are light and little, they do most of the moving." [626]

Professor Lisa Randall says, "Not only our vision, but our other senses— touch, smell, taste, and sound— rely on atomic interactions, which rely in turn on the interactions of electrically charged particles. Touch too, though for more subtle reasons, relies on electromagnetic vibrations and interactions." [627]

Scientists at MIT and Cambridge University have identified an unexpected shared pattern in the collective movement of bacteria and electrons. As billions of bacteria stream through a microfluidic lattice, they synchronize and swim in patterns similar to those of electrons orbiting around atomic nuclei in a magnetic material. The researchers found that by tuning certain dimensions of the microfluidic lattice, they were able to direct billions of microbes to align and swim in the same direction, much the way electrons circulate in the same direction when they create a magnetic field. The researchers also identified a mathematical model that applies to the motions of both bacteria and electrons. The model derives from a general lattice field theory, which is typically used to describe the quantum behavior of electrons in magnetic and electronic materials. [628]

Dr. Rhawn Joseph, a Neuropsychologist says, "The infinite, eternal universe continually recycles energy and mass at both the subatomic and macro-atomic level, thereby destroying and then reassembling atoms, molecules, stars, planets and galaxies... Mass, molecules, atoms, protons, electrons, and elementary particles are continually created and destroyed, and matter and energy, including hydrogen atoms, are continually recycled and recreated...again and again." [629]

Dr. Percy Seymour says, "The electrical interactions between the electrons of one chemical element and those of another chemical element are responsible for the combining of elements to form chemical compounds. The forces that combine all solid materials together are also associated with the behavior of the electrons. Thus, on the everyday level of existence, the solid feel of matter is basically electrical in origin – or more generally, electromagnetic in origin." [630]

Dr. Will Grundy, an Astronomer at Lowell Observatory in Arizona says that material from the Kuiper belt may have delivered organic molecules to Earth. "There's no doubt that complex organic molecules do exist in the outer solar system which were created via energetic radiation acting on simpler molecules...Whether [they] played an important role in seeding life on Earth is not yet known...It's entirely possible that we didn't need that stuff. But it's equally possible that after the Earth had been sterilized and cooked and things cooled down, impactors from the outer solar system [delivered key] astrobiological ingredients." [631]

Brian van der Spuy at Quora.com has calculated that the average human body contains about 1.5 x 10^28 electrons or 15,000,000,000,000,000,000,000,000,000 fundamental electrons. [632]

At any given moment a certain percentage of these electrons will be interacting. If a percentage as low as .01% is surmised then 1,500,000,000,000,000,000,000,000 electrons would be interacting inside and around the human body. This number or any more accurate number of interacting electrons is not to be ignored nor is the complexity based on the number of possibilities.

Astrologer Dr. Bernadette Brady says, "An example of complexity thinking is to consider the Earth's biosphere as a place of complexity. It is a thin film wrapped around the Earth about twenty miles thick...It is also the place where all life on Earth exists...And according to complexity, it is this zone we live our lives in an order of repeating patterns. Similarly the way an individual organism moves through its life-being influenced by an influencing its environment-means that it too, will be involved in these naturally occurring patterns and experience them as upheavals and disturbances, times of ease and times of stress." [633]

Results published in the September 16, 2015 issue of the journal *Nature* say that for the first time direct interspecies electron transport--the movement of electrons from a cell, through the external environment, to another cell type--has been documented in microorganisms in nature. Victoria Orphan, Professor of Geobiology says, "What is really notable is that there are cells that are many cell lengths away from their nearest partner that are still active...This pointed to the possibility that these archaea were directly transferring electrons derived from methane to the outside of the cell, and those electrons were being passed to the bacteria directly. Cultured bacteria use large proteins, called multi-heme cytochromes, on their outer surface that act as conductive "wires" for the transport of electrons. It's really one of the first examples of direct interspecies electron transfer occurring between uncultured microorganisms in the environment. Our hunch is that this is going to be more common than is currently recognized." [635]

Theoretical Physicists at the *Université libre de Bruxelles* have developed a fully time-symmetric formulation of quantum theory which establishes an exact link between asymmetry and the fact that we can remember the past but not the future. The study offers new insights into the concepts of free choice and causality, and suggests that causality need not be considered a fundamental principle of physics. The idea that our choices at the present can influence events in the future but not in the past is reflected in a principle that quantum theorists call causality. According to the principle of causality, the choice of measurement can be correlated with outcomes of measurements in the future only, whereas the outcome of a measurement can be correlated with outcomes of both past and future measurements. The researchers argue that the defining property according to which we interpret the variable describing the measurement is that it can be known before the actual measurement takes place. From this perspective, the principle of causality can be understood as a constraint on the information available about different variables at different times. This constraint is not time-symmetric since both the choice of measurement and the outcome of a measurement can be known a posteriori. This, according to the study, is the essence of the asymmetry implicit in the standard formulation of quantum theory. Dr. Ognyan Oreshkov, the lead author of the study says, "Quantum theory has been formulated based on asymmetric concepts that reflect the fact that we can know the past and are interested in predicting the future. But the concept of probability is independent of time, and from a physics perspective it makes sense to try to formulate the theory in fundamentally symmetric terms." To this end, the authors propose to adopt a new notion of measurement that is not defined only based on variables in the past, but can depend on variables in the future too. Professor Nicolas Cerf, a Co-author of the study and Director of the Centre for Quantum Information and Communication at ULB says, "In the approach we propose, measurements are not interpreted as up to the free choices of agents, but simply describe information about the possible events in different regions of space-time." [636]

Professor Istvan Bokkon at Vision Research Institute and Dr. Vahid Salari, Assistant Professor of Physics, Isfahan University of Technology say that living cells can generate and use electric, magnetic, and electromagnetic waves. Electricity is a basic trait of cells, because all biomolecules are ions which are endowed with high electric dipole moments. When their charges move, an electromagnetic field is generated. Magnetic features can emerge from free radicals, organic molecules with metals or biomagnetites. Different parts of cells, such as DNA and RNA, protein show piezoelectric and semiconductor properties. The electric fields of a wave can couple to the mobile carriers within a semiconductor structure and modify its electronic and elastic properties. In brief, living cells can generate and use electric, magnetic, electromagnetic, and acoustic waves collectively called conformons and convert them from one form to another. [637]

Professor Bokkon and Dr. Salari found that electrons play a very important role in the information flow between organic and inorganic materials in various cells. A functional electric connection can exist between organic and inorganic materials. Bioelectromagnetic forces can regulate the formation of biomagnetites in cells as well. Electromagnetic waves mostly affect the length of cell membranes which are in functional connection with biomagnetites. [638]

Professor Bokkon suggests that the phase of the electron wave depends on the magnetic vector potential, which causes a phase difference and interference between partial waves. Through the Aharonov–Bohm effect, weak geomagnetic fields can have effects on living cellular processes. Biomagnetites can take part in information storage and operating processes in cells. Layers in biomagnetites are shaped by a slow extraction which can be directed via electric and electromagnetic cellular processes. Layers of biomagnetites and phases of non-conductive electrons in the layers are shaped by the current magnetic vector potential field of Earth. Oscillations of electric resistance in biomagnetites and also transport of spins into semiconducting proteins can change conformations of organic molecules, which are in direct connection with biomagnetites. Then, conformational changes can oscillate and these oscillations which appear as conformons generate cellular electromagnetic fields around themselves, which can mediate long-range interactions and also signal-amplifying processes. Electromechanical, electrochemical, or electromagnetic signals could regulate the signal-amplifying processes within and between cells. There are billions of magnetites in the human brain. Professor Joseph Kirschvink, at the California Institute of Technology suggests that while biomagnetites are built up jointly with organic molecules and cellular electromagnetic fields in cells, these biomagnetites can record information of the Earth's magnetic vector. [639]

Sources for Life

[621] 2007, Catling, David C. Astrobiology: *A Very Short Introduction* Oxford University Press, Kindle Edition (L 630)
[622] 2007, Catling, David C. *Astrobiology: A Very Short Introduction* Oxford University Press, USA. Kindle Edition (L 630)
[623] 2007, Catling, David C. *Astrobiology: A Very Short Introduction* Oxford University Press, Kindle Edition (L 339)
[624] 2007, Catling, David C. *Astrobiology: A Very Short Introduction* Oxford University Press, USA. Kindle Edition. (L 702)
[624a] 1992, 2011, Hickey, Isabel M. *Astrology, A Cosmic Science: The Classic Work on Spiritual Astrology* Kindle Edition. (L 20) SCB Distributors.
[624b] 1992, 2011, Hickey, Isabel M. *Astrology, A Cosmic Science: The Classic Work on Spiritual Astrology* Kindle Edition.(L 139) SCB Distributors.
[625] 2014, New Scientist
https://www.newscientist.com/article/dn25894-meet-the-electric-life-forms-that-live-on-pure-energy
[626] 2015, Astrologer John Rutherford, by permission October 2015
[627] 2015, Randall, Lisa, *Dark Matter and the Dinosaurs: The Astounding Interconnectedness of the Universe* Kindle Edition, Harper Collins. (L 433)
[628] 2016, Massachusetts Institute of Technology
http://www.sciencedaily.com/releases/2016/01/160105133132.htm?utm_source=feedburner&utm_medium=email&utm_campaign=Feed%3A+sciencedaily%2Ftop_news%2Ftop_science+%28ScienceDaily%3A+Top+Science+News%29
[629] 2010, Journal of Cosmology
http://journalofcosmology.com/Cosmology5.html
[630] 2008, Seymour, Percy *Dark Matters: Unifying Matter, Dark Matter, Dark Energy, and the Universal Grid* (L 96) Kindle Edition.
[631] 2015, The Telegraph, U.K.
http://www.telegraph.co.uk/news/science/space/11335524/NASA-probe-to-arrive-at-Pluto-carrying-ashes-of-Clyde-Tombaugh.html
[632] 2015, Calculations by Brian van der Spuy, January 24, 2015
www.Quora.com

[633] 2014, Brady, Bernadette, *Cosmos, Chaosmos and Astrology*, Page 8

[635] 2015, California Institute of Technology
http://www.sciencedaily.com/releases/2015/09/150918180315.htm?utm_source=feedburner&utm_medium=email&utm_campaign=Feed%3A+sciencedaily%2Ftop_news%2Ftop_science+%28ScienceDaily%3A+Top+Science+News%29

[636] 2015, Libre de Bruxelles, Université
http://www.sciencedaily.com/releases/2015/07/150728091946.htm?utm_source=feedburner&utm_medium=email&utm_campaign=Feed%3A+sciencedaily%2Ftop_news%2Ftop_science+%28ScienceDaily%3A+Top+Science+News%29

[637] 2009, Original Paper
http://www.ncbi.nlm.nih.gov/pmc/articles/PMC2791810/pdf/10867_2009_Article_9173.pdf

[638] 2009, Original Paper
http://www.ncbi.nlm.nih.gov/pmc/articles/PMC2791810/pdf/10867_2009_Article_9173.pdf

[639] 2009, Original Paper
http://www.ncbi.nlm.nih.gov/pmc/articles/PMC2791810/pdf/10867_2009_Article_9173.pdf

Why Not

Neutrons

CPI Theory has singled out the electron at the exclusion of the neutron so why not the neutron? The simple answer is that neutrons are not alive or activated without the electron but the electron is active as a free electron without the neutron. The neutron is vital for life but only after the electron has intervened.

Professor Philip Ball says, "The configuration of electrons is the same for all isotopes of an element – adding extra neutrons to a nucleus has essentially no effect on the electrons." [640] A small mass difference, is the reason why free neutrons decay on average after around ten minutes, while protons, the unchanging building blocks of matter, remain stable for a practically unlimited period. [641]

Protons

CPI Theory has singled out the electron at the exclusion of the proton so why not the proton? The simple answer is that protons are not alive or activated without the electron but the electron is active as a free electron without the proton. The proton is vital for life but only after the electron has intervened.

A proton is approximately 1836 times heavier than an electron and in addition to being subject to the electromagnetic force it is also subject to the so-called strong interaction, which is responsible for the structure and the cohesion of atomic nuclei. [642]

Professor Andrew King says, "The number of protons in the nucleus is exactly equal to the number of electrons in an uncharged atom of that element, and so also equal to the atomic number." [643] The atomic number is also equal to the number of electrons and although the number of protons and electrons are equal it is the electron that is the guiding force.

Photons

CPI Theory has singled out the electron at the exclusion of the photon so why not the photon? The simple answer is that photons are not alive or activated without the electron but the electron is active as a free electron without the photon. The photon is a visualization of the electron in action.

Professor Stephen J. Blundell says, "The oscillations of electricity and magnetism in a beam of light are governed by Maxwell's four beautiful equations, operating together like the cogs, wheels, and spindles inside an intricate machine, each playing a role to keep the whole wonderful mechanism in perfect harmony." [644] Dr. Rodney A. Brooks says, "The energy of a photon is proportional to its frequency of oscillation." [645]

Professor Peter Lodahl, at the Niels Bohr Institute at the University of Copenhagen says, "The smallest component of light is a photon and photons are very well suited for carrying information. A quantum circuit based on photons could contain far more information than is possible with current computer technology and the information could not be intercepted en route." Researchers have developed a photonic chip, in which a light source, a so-called quantum dot, is embedded. By shining light on the quantum dot using a laser, its electrons are excited, which then jump from one orbit to another and thus emit a single photon at a time. Light is normally emitted in all directions, but the photonic chip is constructed so that all of the photons are sent out through a photonic channel. The problem is that the photons are sent in both directions in the photonic channel and this limits the efficiency of the light source. Professor Lodahl says, "In our work...we have now developed a new photonic channel where we can control the photons so that they are only sent in one direction. It is a fundamental new discovery, that you can get the emission of light in a photonic chip to take place in a manner not previously thought possible." Dr. Immo Söllner and Dr. Sahand Mahmoodian at the Niels Bohr Institute at the University of Copenhagen say that they use laser light to excite the quantum dot's electrons, which jump from one orbit to another and thereby emit a single photon. By controlling the spin of the electrons with a magnetic field, you can get an entirely different light emission. A photon emitted from a quantum dot with an electron spin-down chooses one direction, while the photon from a quantum dot with an electron spin-up chooses the opposite direction. The direction of the light emission depends on the spin of the quantum dots. [646]

When two photons collide we get a positron and an electron and when an electron and a positron collide we generate two photons. [647] The electrons can absorb these photons and lift themselves to more energetic orbits. Eventually, the electron drops back from its higher-energy orbit, and re-emits the photon. [648] When an electron wants to create light it absorbs a positron and emits a pair of photons. When an electron wants invisibility it releases the positron.

In a conversation with Astrologer John Rutherford:

JR...Electrons and photons are separate entities with very different qualities.

TW... Yes, that is what science says but I believe that much of what is going on is either electron controlled or is the electron in another phase of action.

JR...A photon is produced when an electron loses energy...

TW...Yes but, when an electron looses energy the electron can be seen as light thus a photon is an electron phase and not a separate entity.

JR...and one is absorbed when an electron absorbs energy.

TW...Yes again, as the electron absorbs energy the electron can be seen as light. An electron signals via light that it is in an energy transition phase.

JR...Therefore, a photon is a separate energy packet, not an electron phase.

TW...Ok, if I am right what kind of energy does a photon have but an electron energy? Since light is by definition an electromagnetic wave then it is already half electron due to the electro and it is the electron that causes the magnetic. [649]

Professor Richard P. Feynman said, "When a photon is absorbed by an electron, the electron continues on a bit, and a new photon comes out. This process is called the scattering of light [650] ...Even more strange is the possibility that the electron emits a photon, then travels backwards in time to absorb a photon, and then proceeds forwards in time again [651] ...As the electron and the nucleus are exchanging photons, a photon comes from outside the atom, hits the electron and is absorbed then a new photon is emitted." [652] Appropriately William Shakespeare has said, "...what's past is prologue..." [653]

Dr. Rodney A. Brooks says, "...if a photon is an entity that lives and dies as a unit, field collapse must occur [654] ...Field collapse is when a photon is absorbed, no matter how spread out it may be, all its energy is deposited into the absorber [655] ...I am asking you to believe that this field quantum, spread out as it may be, suddenly disappears into a tiny absorbing atom [656] ...In Quantum Field Theory the photon is a spread-out field, and the particle-like behavior occurs because each photon, or quantum of field, is absorbed as a unit. When a spread-out photon is absorbed by an atom, the entire field vanishes and all its energy is deposited into the atom. This process is known as field collapse... [657]

Professor Richard P. Feynman said, "The first basic action, the first basic law of physics— a photon goes from point to point [658] ...The second action fundamental to quantum electrodynamics is that an electron goes from point A to point B in space-time [659] ...However, you might be interested to know that the formula for a photon going from place to place in space-time is the same as an electron going from place to place... [660]

Another conversation with Astrologer John Rutherford:
JR....Electron orbits describe energy levels of the electron and are fixed for each atom. Atoms of different atomic weights have differing energy levels for their respective electron shells.
TW...Is it not true that atomic weight is caused by accumulating electrons thus the energy levels are really electron accumulation levels?
JR...Changing orbit levels produce photons, light, but the electron still remains unlit, invisible, until detected by other means.

TW...If the electron remains invisible as it changes orbital how can it be clearly said that the photon is not electron visualization? [661]

Plasma

CPI Theory has singled out the electron at the exclusion of plasma so why not plasma? The simple answer is that in solar plasma a proton is surrounded and transported by electrons. The electrons seek out the destination for the plasma namely an ionosphere where the transporting electrons release their proton cargo. In the ionosphere the plasma-less proton is surrounded by free electrons that facilitate an atom's construction and thus produce a chemical element. As the atom is constructed the proton activates and becomes part of an active nucleus. Through the guidance of the electron via the interaction of the outer shell electrons one atom merges with other atoms to become molecules and cells. Plasma is not alive or activated without the electron but the electron is active as a free electron without the proton. Plasma is vital for proton transportation but only after the electron has intervened.

Dr. Syun-Ichi Akasofu says, "Chapman and Ferraro (1931) showed that solar gas must be treated as plasma not as a cloud of solitary particles because protons and electrons in the stream are strongly coupled in motion as they flow around the Earth's dipole field." [662]

Earth's magnetosphere, is filled with plasma that is created by the atmosphere being ionised by sunlight. The innermost layer of the magnetosphere is the ionosphere, and above that is the plasmasphere. Both are embedded with a variety of strangely shaped plasma structures including tubes. Cleo Loi an

Astrophysicist at the ARC Centre of Excellence for All-sky Astrophysics and School of Physics at the University of Sydney in Australia says, "We measured their [tube structures] position to be about 600 kilometres above the ground, in the upper ionosphere, and they appear to be continuing upwards into the plasmasphere. This is around where the neutral atmosphere ends, and we are transitioning to the plasma of outer space...We saw a striking pattern in the sky where stripes of high-density plasma neatly alternated with stripes of low-density plasma. This pattern drifted slowly and aligned beautifully with the Earth's magnetic field lines, like aurora..." [663]

Neutrinos

CPI Theory has singled out the electron at the exclusion of the neutrino so why not the neutrino? The simple answer is that the neutrino is really the electron separated from its charge. The neutrino is a phase of the electron in action acting in a non-electromagnetic way. A neutrino is formed when atoms join and release creating energy.

University of California, Irvine says, "Neutrinos are similar to the electron but neutrinos do not carry electric charge. Because neutrinos are electrically neutral, they are not affected by the electromagnetic forces which act on electrons. Three types of neutrinos are known. Each type of neutrino is related to a charged particle. The electron neutrino is associated with the electron, and two other neutrinos are associated with heavier versions of the electron called the muon and the tau." [664]

At almost at the speed of light, thousands of millions of neutrinos from interplanetary space and beyond pass through Earth. [665]

Sources for Why Not

[640] 2004, Ball, Philip, *The Elements: A Very Short Introduction* Oxford University Press. Kindle Edition. (L 1816)
[641] 2015, Forschungszentrum Juelich
http://www.sciencedaily.com/releases/2015/03/150326151607.htm?
utm_source=feedburner&utm_medium=email&utm_campaign=Feed
%3A+sciencedaily%2Ftop_news%2Ftop_science+%28ScienceDaily
%3A+Top+Science+News%29
[642] 2014, Physikalisch-Technische Bundesanstalt (PTB)
http://www.sciencedaily.com/releases/2014/11/141118072744.htm?
utm_source=feedburner&utm_medium=email&utm_campaign=Feed
%3A+sciencedaily%2Ftop_news%2Ftop_science+%28ScienceDaily
%3A+Top+Science+News%29
[643] 2012, King, Andrew, *Stars: A Very Short Introduction* Oxford University Press. Kindle Edition. (L 613) (p. 33)
[644] 2012, Blundell, Stephen J. *Magnetism: A Very Short Introduction* Oxford University Press. Kindle Edition. (L 842)
[645] 2010, Brooks, Rodney A. *Fields of Color: The theory that escaped Einstein* Epic Publications. Kindle Edition. (L 1184)
[646] 2015, University of Copenhagen - Niels Bohr Institute
http://www.sciencedaily.com/releases/2015/07/150727120220.htm?
utm_source=feedburner&utm_medium=email&utm_campaign=Feed
%3A+sciencedaily%2Ftop_news%2Ftop_science+%28ScienceDaily
%3A+Top+Science+News%29
[647] 2011, Hall, Alan *Electron* (L 171) Kindle Edition.
[648] 2012, King, Andrew, *Stars: A Very Short Introduction* (L 571) Oxford University Press. Kindle Edition.
[649] 2015, Astrologer John Rutherford by permission October 2015
[650] 2014, Feynman, Richard P. *QED: The Strange Theory of Light and Matter* Princeton University Press Kindle Edition. (L1591)

[651] 2014, Feynman, Richard P. *QED: The Strange Theory of Light and Matter* Princeton University Press Kindle Edition. (L1591)

[652] 2014, Feynman, Richard P. *QED: The Strange Theory of Light and Matter* Princeton University Press Kindle Edition. (L1641)

[653] 1610, William Shakespeare, *The Tempest, Act II, scene i*

[654] 2010, Brooks, Rodney A. *Fields of Color: The theory that escaped Einstein* Epic Publications. Kindle Edition. (L 1129)

[655] 2010, Brooks, Rodney A. *Fields of Color: The theory that escaped Einstein* Epic Publications. Kindle Edition. (L 1187)

[656] 2010, Brooks, Rodney A. *Fields of Color: The theory that escaped Einstein* Epic Publications. Kindle Edition. (L 1128)

[657] 2010, Brooks, Rodney A. *Fields of Color: The theory that escaped Einstein* Epic Publications. Kindle Edition. (L 1121)

[658] 2014, Feynman, Richard P. *QED: The Strange Theory of Light and Matter* Princeton University Press Kindle Edition (L 1484)

[659] 2014, Feynman, Richard P. *QED: The Strange Theory of Light and Matter* Princeton University Press. Kindle Edition (p. 90)

[660] 2014, Feynman, Richard P. *QED: The Strange Theory of Light and Matter* Princeton University Press Kindle Edition (L 1498)

[661] 2015, Astrologer John Rutherford by permission October 2015

[662] 2007, Akasofu, Syun-Ichi, *Exploring the Secrets of the Aurora* Kindle. (L 361)

[663] 2015, University of Sydney
http://www.sciencedaily.com/releases/2015/06/150601092156.htm

[664] 2015, University of California Irvine
http://www.ps.uci.edu/~superk/neutrino.html

[665] 2014, Universidad de Barcelona
http://www.sciencedaily.com/releases/2014/07/140722091326.htm?utm_source=feedburner&utm_medium=email&utm_campaign=Feed%3A+sciencedaily%2Ftop_news%2Ftop_science+%28ScienceDaily%3A+Top+Science+News%29

Summary

To summarize the Abstract

- *CPI Theory Part Three advances the discussion of continuous planetary interaction within our solar system*

- *at all times I will attempt to stay focused on continuous planetary interaction*

To summarize the Introduction

- *aurora seen flowing from an auroral oval is the visualized radiation of the electromagnetic interaction between a planet and interplanetary space*

- *there are two known types of magnetized planets in our solar system and both types can have an auroral oval and radiate aurora*

To summarize Magnetic Planets

- *all planets in our solar system have either a magnetic core or a magnetic field and sometimes both*

- *magnetic fields produce the magnetosphere*

- *a magnetic field encapsulates an ionosphere*

- *in order to have an ionosphere a planet must be magnetically active with either a magnetic core dynamo or a magnetic field*

- *it is understood that to have an auroral oval a planet must have a magnetic field*

- *since the magnetic field of Earth changes over time then possibly the magnetic fields of other planets also change*

- *from an astrological point of view any change of planetary magnetic strength could account for changes in the type of astrological influence*

To summarize Planetary and Interplanetary Fields

- *fields are a property of space, not a separate substance in space*

- *fields are conditions of space itself, considered apart from any matter that may be in it*

- *the primary elements of reality are not individual particles, but underlying fields*

- *all electrons are but excitations of an underlying field, the electron field, which fills all space and time*

- *a quantum of the electromagnetic field can be emitted from an atom in the Sun, travel through millions of miles of space, spreading out as it goes, and then interact as a single unit with an atom in your eye*

- *fields can have mass and charge*

- *all physical properties – spin, mass, charge, energy, etc. – must be properties of fields*

- *electric and magnetic fields are two aspects of the way charged objects exert forces on each other*

- *a charged particle in motion becomes an electric current and creates a magnetic field where a nearby moving charge will feel a force because of this magnetic field*

- *changes in an electric field will produce changes in a magnetic field, and vice versa, and a self-sustaining wave of varying electric and magnetic fields will propagate off into space*

To summarize Spin and Orbit

- *the spinning motion of the electron gives rise to a magnetic field*

- *in an atom, magnetism arises from the spin and orbital momentum of its electrons*

- *the spinning direction of the electrons defines the direction of the magnetization in a material*

- *an increasing magnetic field allows energy to escape as the electron spins realign*

- *each orbit corresponds to a precise and distinct energy for the electron to occupy*

- *the energy lost by the electron while jumping from the outer to the inner orbit is emitted as light by the atom*

- *the energy of the emitted light makes up for the difference between the energies in the two orbits*

To summarize Waves

- *an important property of waves is interference*

- *interference occurs when two fields converge at a point in space and either reinforce or cancel each other, depending on the direction of their forces*

- *electrons are elementary particles that are indivisible and unbreakable*

- *Schrödinger developed an equation where the wave shape of an electron was defined*

- *Paul Dirac said that the entire science of chemistry can be derived from the Schrödinger equation*

- *according to Physicist Niels Bohr a particle and a wave were complimentary concepts giving different representations of the same object*

- *the electron goes in any direction at any speed, forward or backward in time*

- *whenever the edge of Earth's magnetosphere, the magnetopause, gets rattled it will create waves that propagate everywhere in the magnetosphere*

- *whistler-mode chorus waves have important roles in both acceleration and loss of energetic radiation belt electrons*

- *chorus wave activity is dependent on geomagnetic activity and occurs over a wide range of geospace*

- *great archetypal wave forms that emerge, crest, and then surge through the collective or individual psyche and lifeworld*

To summarize Electron Flow

- *when electrons exit a planet they rise out of the ionosphere passing through an auroral oval*

- *the electron is necessary for everything*

- *the electron is not only necessary but I think it has a purpose and destiny*

- *where you find energy interaction you will find the electron and should find astrological signatures*

- *whenever electrons are disturbed we see the advent of electromagnetism*

- *the Earth's north magnetic pole is a negative pole, because the positive north-seeking end of a compass needle is attracted toward it*

- *electrons come into both poles they just curve in differing directions*

- *electrons on the other hand, do flow since they can jump from atom to atom so negative charges can flow*

- *electron flow can be quite clear if you just follow the electron and accept that the electron is guiding the flow and subsequent interaction*

To summarize Ions, Ionization Energy & the Ionosphere

- *an ion is an atom with an electric charge*

- *an ion is an atom or molecule which has gained or lost one or more of its valence electrons*

- *every ion has a charge, either positive or negative*

- *an ion is an atom or group of atoms in which the number of electrons is different from the number of protons*

- *ions are still atoms*

- *compare the astrological qualities cardinal, fixed and mutable with the three types of atom, negative, neutral and positive*

- *an entering electron is a cardinal electron*

- *a vacating electron is a mutable electron*

- *an electron that is not entering or vacating is a fixed electron*

- *the higher the temperature the more active is the process of mutable electrons becoming cardinal electrons*

- *mutable electrons become cardinal electrons by adding one electron*

- *a mutable electron realigns its spin when it pairs with another mutable electron and a cardinal electron is created*

- *ionization energy is the quantity of energy that an atom must absorb to discharge an electron*

- *the further away an electron is from the nucleus, the easier it is for it to be expelled*

- *the ionosphere is so named because it is the area where ions and ionization energy occur*

- *when electromagnetically attracted electrons arrive at Earth's poles they enter the auroral oval and initiate changes within the ionosphere*

To summarize Magnetotail

- *the magnetotail is simply the tail of the ionosphere*

- *flux transfer events take place roughly every 8 minutes at Earth*

- *reconnection at the Sun propels plasma clouds toward Earth where additional reconnection events then sparks auroras*

To summarize Atmosphere

- *Earth's atmosphere is the biochemical mixing ground located between our planet's surface and the outer edges of the magnetosphere*

- *encased within the magnetosphere, the atmosphere is constantly reacting to electromagnetic changes brought about by electrons entering the magnetosphere through the auroral oval*

- *there was an enormous decrease in noctilucent cloud frequency 15 to 25 days after the summer solstice*

- *noctilucent clouds are a sensitive indicator of long range teleconnectors in Earth's atmosphere, which link weather and climate across hemispheres*

- *the first observational study to show the link between polar mesospheric clouds and planetary wave activity, which is the driver of variability in the winds*

- *the first study to suggest a direct link between the polar summer mesosphere and the surface in the winter hemisphere*

- *during the summer polar mesospheric clouds occur 50 miles above the surface of the polar regions*

To summarize Lightning

- *lightning is another expression of the electron*

- *lightning is found at Earth's exosphere and in Earth's atmosphere near ground level*

- *lightning in the lower atmosphere is positive and negative interaction between water crystals in the air and the Earth caused by temperature changes, the same activity that excites electrons*

- *in the lower atmosphere lightning involves a separation of charge*

- *in the upper atmosphere lightning involves a freezing process*

- *excess electrons on the bottom of the cloud begin a journey through the conducting air to the ground at speeds up to 60 miles per second*

- *as the step leader grows it provides the roadway between cloud and Earth along which the lightning bolt will eventually travel*

- *as many as a billion trillion electrons can transverse this path in less than a millisecond*

- *the expansion of the air creates a shockwave that we observe as thunder*

- *wherever you are, if it's 8 a.m., it's time for the most powerful lightning strikes of the day*

- *lightning strikes the Earth more than 4 million times a day*

- *the UK experienced around 50% more lightning strikes when the Earth's magnetic field was bent by the Sun's own magnetic field*

- *an electron emits light as a by-product of one of its transfer processes*

- *lightning is one form of the electron facilitating transfer*

To summarize Temperature and Superconductivity

- *electrons move in response to either an electric field or a temperature gradient*

- *as electrons in an atom encounter increasing temperature they are raised to ever higher energy levels*

- *the higher the temperature the more active are the entering electrons*

- *at low temperature, fewer vibrations mean less scattering of electrons and the resistance falls*

- *for superconductivity to occur, the material has to be cooled to a very low temperature*

- *above the critical temperature, the net interaction between two electrons is repulsive*

- *below the critical temperature the overall interaction between two electrons becomes very slightly attractive*

- *the classical elements are familiar representatives of the different physical states that matter can adopt*

To summarize Aurora

- *an auroral oval is caused by the combination of the solar wind and a planet with a magnetic field*

- *the visual effects of an auroral oval and the energy excitement within an ionosphere are caused by electrons*

- *aurora can come in different colors due to the atmospheric particles interacting in the ionosphere*

- *the colours relate to the degree of ionization of the elements available, i.e. the strength of the storm*

- *in a typical auroral display the light is a mixture of many colors*

- *red aurora is associated with intense solar activity and heating of the upper atmosphere from a large influx of low-energy electrons*

- *black aurora occurs where there are holes in the ionosphere where particles are shooting upwards into interplanetary space*

- *ordinary aurora are associated with downward-flowing electrons bombarding the atmosphere, the black ones are associated with electrons being sucked out from the atmospheres into interplanetary space leaving deep cavities in the upper ionosphere*

- *a drop in low-energy electrons, long thought to have little or no effect corresponds with especially fast changes in the shape and structure of pulsating auroras*

- *a proton aurora is a form of aurora caused by heavy solar ions striking Earth's upper atmosphere*

- *the astrological zenith is the part of the zodiacal chart where the auroral oval empties its electrons due to solar pressure*

- *it could be said that the zenith is the center of the inflow radiation into the zodiacal chart while the eastern horizon is the entrance into the zodiacal chart and into the natural where we live*

- *the hemispheric power is an energy estimate of all electrons precipitating into a hemisphere*

- *the sum of the hemispheric power in both northern and southern hemispheres is the auroral electron power*

To summarize The Sun and the Heliosphere

- *there is still a lot of uncertainty as to how the Sun affects the climate*

- *a study from Lund University in Sweden suggests that direct solar energy is not the most important factor, but rather indirect effects on atmospheric circulation*

- *solar minimum and maximum are opposite extremes of the solar rhythm*

- *the heliosphere encapsulates our solar system*

- *the heliosphere is the magnetic bubble inflated from the inside by the high-speed solar wind blowing out from the Sun*

- *the bow shock pushes ahead through the interstellar medium as the heliosphere plows through the galaxy*

- *the heliospheric current sheet is also sometimes called the interplanetary current sheet*

- *the interplanetary current sheet reaches Earth's ring current where electrons in the solar wind are accelerated into the auroral oval*

- *the Sun's magnetic field controls the large-scale shape of the heliosphere much more than expected*

To summarize Inner Planets

- *Mercury's magnetic field appears to be generated by an active dynamo in the planet's core*

- *Mercury's magnetic field is similar to Earth's, but much weaker*

- *Venera 4 found the magnetic field on Venus to be induced by an interaction between the ionosphere and the solar wind*

- *the solar wind is interacting directly with Venus' outer atmosphere where ions of hydrogen and oxygen are being created*

- *European Space Agency scientists reported that the ionosphere of Venus streams outwards in a manner similar to the ion tail seen streaming from a comet under similar conditions*

- *plasma waves buffeting Earth's radiation belts are responsible for scattering charged particles into the atmosphere*
- *the Van Allen radiation belts periodically swell and shrink in response to incoming energy disturbances from the Sun*

- *2014 findings suggest that the interior of the Moon has not yet cooled and hardened, and also that it is still being warmed by the effect of the Earth on the Moon*

- *airless bodies experience a suite of processes, collectively described as space weathering, that slowly alter the composition of the surface*

- *2015 data show that, on Mars, aurora in the upper atmosphere glows blue depending on the activity of the Sun*

- *Mars' upper atmosphere is more like Earth's than previously thought*

- *NASA's MAVEN spacecraft detected evidence of widespread auroras in Mars' northern hemisphere*

- *magnetic fields in the solar wind drape across Mars, even into the atmosphere, and the charged particles just follow those field lines down into the atmosphere*

- *comet Siding Spring caused an intense meteor shower on Mars and added a new layer of ions to the ionosphere*

- *the data are consistent with a few tons of comet dust being deposited in the atmosphere of Mars*

- *Mars Express observed a huge increase in the density of electrons following Siding Spring's close approach*

To summarize Asteroids and Comets

- *the asteroid belt is virtually empty space*

- *interparticle cohesive forces must be holding it together*

- *the Rosetta spacecraft has been watching the early stages of how a magnetosphere forms around comet 67P Churyumov-Gerasimenko*

- *the comet atmosphere appears to be very unevenly distributed around the nucleus*

- *Rosetta spacecraft in orbit around comet 67P/Churyumov-Gerasimenko detected 16 organic compounds as it descended toward and then bounced across the comet's surface*

- *the team found 21 different organic molecules in gas from the comet, including ethyl alcohol and glycolaldehyde, a simple sugar*

To summarize Outer Planets

- *the energetic particles coming from Jupiter's magnetosphere create bright auroral ovals, which encircle the poles of Jupiter*

- *Jupiter has powerful storms, often accompanied by lightning strikes*

- *computer simulations show that Jupiter ejected more water-rich material than it scattered inward*

- *Hubble has tracked Jupiter's moon Ganymede and has shown how the magnetic field draws in and excites space particles, generating a glow of ultraviolet light around the north and south poles*

- *Saturn's magnetosphere, like Earth's, produces aurora*

- *Saturn's planetary magnetic field, which is weaker than Earth's, but has a magnetic field 580 times that of Earth due to Saturn's larger size*

- *Saturn's magnetic field strength is around one-twentieth the strength of Jupiter's*

- *photochemical cycle is modulated by Saturn's annual seasonal cycle*

- *since early 2005 scientists have been tracking lightning on Saturn*

- *the moon Titan orbits within the outer part of Saturn's magnetosphere and contributes plasma from the ionized particles in Titan's outer atmosphere*

- *Titan, Saturn's largest moon is larger than the planet Mercury and is the only moon in the solar system to have a substantial atmosphere*

- *interactions between Titan's atmosphere, and the solar magnetic field and radiation, create a wind of hydrocarbons and nitriles that blow away from Titan's polar regions into interplanetary space*

- *in 2011 astronomers caught the first views of auroras on Uranus*

- *the intensity of the magnetic field at Uranus' surface is roughly comparable to that of Earth's*

- *the gas-giant Uranus is also considered by NASA to be an ice giant*

- *Neptune, which is Uranus' near twin in size and composition, radiates 2.61 times as much energy into space as it receives from the Sun*

- *Uranus radiates hardly any excess heat at all*

- *Voyager detected auroras, similar to aurora on Earth, in Neptune's atmosphere*

- *due to Neptune's complex magnetic field the auroras are an extremely complicated processes that occurs over wide regions of Neptune not just near the magnetic poles*

- *Neptune has a magnetosphere*

- *Neptune's orbit has a profound impact on the Kuiper belt a ring of small icy worlds, similar to the asteroid belt but far larger*

- *Pluto is the leading representative of the Kuiper Belt*

- *Pluto has a wispy but enormous atmosphere*

- *Pluto's nitrogen-rich atmosphere is quite extended*

- *Charon is one of the largest objects in the Kuiper Belt*

- *Charon orbits quite close to Pluto about 12,000 miles away*

To summarize Life

- *how life arose is unknown*

- *you eat sugars that have excess electrons, and you breathe in oxygen that willingly takes them*

- *electrons must flow in order for energy to be gained*

- *all life is about the flow of charged particles*

- *life is the continuous flow of ionic charges*

- *the electrical interactions between the electrons of one chemical element and those of another chemical element are responsible for the combining of elements to form chemical compounds*

- *the average human body contains about 1.5 x 10^28 electrons or 15,000,000,000,000,000,000,000,000,000 fundamental electrons*
- *a thin film wrapped around the Earth about twenty miles thick*

- *one of the first examples of direct interspecies electron transfer occurring between uncultured microorganisms in the environment*

- *measurements are not interpreted as up to the free choices of agents, but simply describe information about the possible events in different regions of space-time*

- *the electric fields of a wave can couple to the mobile carriers within a semiconductor structure and modify its electronic and elastic properties*

- *a functional electric connection can exist between organic and inorganic materials*

- *the phase of the electron wave depends on the magnetic vector potential*

- *biomagnetites can record information of the Earth's magnetic vector*

To summarize Why Not

- *the neutron is vital for life but only after the electron has intervened*

- *the proton is vital for life but only after the electron has intervened*

- *the photon is a visualization of the electron in action*

- *the energy of a photon is proportional to its frequency of oscillation*

- *photons are very well suited for carrying information*

- *when an electron wants to create light it absorbs a positron and emits a pair of photons*

- *when an electron wants invisibility it releases the positron*

- *if a photon is an entity that lives and dies as a unit, field collapse must occur*

- *a photon goes from point to point*

- *an electron goes from point A to point B in space-time*

- *electron orbits describe energy levels of the electron and are fixed for each atom*

- *changing orbit levels produce photons*

- *electrons seek out the destination for solar plasma namely an ionosphere*

- *transporting electrons release their proton cargo in the ionosphere*

- *plasma is vital for proton transportation but only after the electron has intervened*

- *the neutrino is a phase of the electron in action acting in a non-electromagnetic way*

- *a neutrino is formed when atoms join and release creating energy*

Conclusions

Continuous Planetary Interaction Theory solidifies my understanding that there is continuous planetary interaction with life on Earth. Across human history astrology has always affirmed that the planets in the solar system interact with us.

It seems that most sciences are talking about the same thing just taking different views from different perspectives. Professor Russell Stannard says, "Different observers can arrive at different conclusions." [666] Professor Richard P. Feynman said, "A possibility is that things look similar because they are aspects of the same thing, some larger picture underneath, from which things can be broken into parts that look different." [667] Different names but it is still the electron interacting and changing.

Professor Itzhak Bars at USC Dornsife College of Letters, Arts and Sciences says, "Quantum mechanics is often counterintuitive, allowing for particles to be in two places at once...It has become an invaluable and accurate framework for understanding the interactions of matter and energy at small distances. Quantum mechanics is extremely successful as a model for how things work on small scales, but it contains a big mystery: *the unexplained foundational quantum commutation rules that predict uncertainty in the position and momentum of every point in the universe.* The commutation rules don't have an explanation from a more fundamental perspective...clearly the rules are correct, but they beg for an explanation of their origins in some physical phenomena that are even deeper..." [668]

Astrology offers a means to understand planetary interaction from a personal or subjective view. The planets and other bodies of our solar system are continuously in motion touching and interacting with everything we know.

Astrologer Bruce Scofield, PhD., says, "I think the CPI hypothesis has outlined a possible mechanism behind at least some astrological phenomena - and its possibilities have been presented in an organized and logical way. It now needs some experimental design and testing." [669]

Robin Armstrong, President of the RASA School of Astrology says, "CPI Theory is a really insightful treatise. I found it filled with useful information and associations. It sets a new standard of perception and deduction in astrology, or at least a new refreshing approach." [670]

A few questions and answers;

Does the history of the electron give it a unique quality? It should.

Does an electron from another planet convey the possibilities of its last location? I think you will find science says yes.

Is the electron influenced by the path it has taken? It should be.
Is every electron unique? Possibly.

Is the atom the end result of the experience of the electron? It is one end result.

Does the electron have a choice to be a lone electron around an atom or is it the electrons destiny to join with other electrons orbiting an atomic nucleus and be part of a group? The collective over the individual.

When an electron leaves a group to be free what is its destiny? The individual over the collective.

What does the electron need? To be continued.

What does the electron want? To be continued.

In places I have personalized the character of the electron but isn't that the point of continuous planetary interaction and its astrological implications. Robin Armstrong, President of the RASA School of Astrology says, "The character of the electron and its primary role in understanding continuous planetary interaction could benefit greatly by applying astrological insights and implications to the cycles. Astrology being the study of relationships and cycles can open many doors or comprehension to what is scientifically known about the electron and life itself." [671]

Have I proven CPI Theory? Time will decide that. However, I have painted a picture of a series of facts that are hard to deny. I believe that a reasonable person thinking reasonably would come to the conclusion that CPI Theory has identified many facts that point to continuous planetary interaction.

Keep on track, *Continuous Planetary Interaction*.

Sources for Conclusions

[666] 2008, Stannard, Russell, *Relativity: A Very Short Introduction* Oxford University Press, Kindle Edition. (L 465)
[667] 2014, Feynman, Richard P. *QED: The Strange Theory of Light and Matter* Kindle Edition Princeton University Press (L 2393)
[668] 2014

http://www.sciencedaily.com/releases/2014/11/141103142326.htm?
utm_source=feedburner&utm_medium=email&utm_campaign=Feed
%3A+sciencedaily%2Ftop_news%2Ftop_science+%28ScienceDaily
%3A+Top+Science+News%29
[669] 2015, Astrologer Bruce Scofield by Permission December
2015. http://www.onereed.com
[670] 2015, Armstrong, Robin, President of the RASA School of
Astrology by permission December 2015 http://www.rasa.ws/
[671] 2015, Armstrong, Robin, President of the RASA School of
Astrology by permission November 2015 http://www.rasa.ws/

Acknowledgements for Part Three

As I have worked my way through CPI Theory, four astrologers have always been willing to lend their expertise. When I needed feedback one of them would rise to the occasion. In CPI Theory Part Three Astrologer John Rutherford and I had numerous one hour telephone conversations as we together read through and discussed each written page and what was being said. Robin Armstrong, President, RASA School of Astrology stepped up to the plate with an excellent written critique that we then refined through several emails. Astrologer Robert Currey, who in January 2016 was given a Lifetime Achievement Award at the 26th Indian Astrological Conference still found time to give encouragement and insight. Astrologer Bruce Scofield, PhD., always offered sage advice and comments. Any mistakes are mine.

Bibliography

Akasofu, Syun-Ichi. *Exploring the Secrets of the Aurora*, Astrophysics and Space Science Library. 2007.
Akasofu, Syun-Ichi. *The Northern Lights*, Alaska Northwest Books, 2009.
Allen, Terence; Cowling, Graham. *The Cell: A Very Short Introduction*, Oxford University Press. Kindle Edition, 2011.
Atkins, Peter. *The Laws of Thermodynamics: A Very Short Introduction,* Oxford University Press. Kindle Edition, 2010.

Ball, Philip. *The Elements: A Very Short Introduction*, Oxford University Press. Kindle Edition, 2004.
Blundell, Stephen J. *Magnetism: A Very Short Introduction*, Oxford University, Kindle Edition, 2012.
Blundell, Stephen J. *Superconductivity: A Very Short Introduction,* Oxford University Press. Kindle Edition, 2009.
Brooks, Rodney A. *Fields of Color: The theory that escaped Einstein,* Epic Publications. Kindle Edition, 2010.
Bryson, George. *Northern Lights, The Science, Myth and Wonder of Aurora Borealis*, 2001

Catling, David C. *Astrobiology: A Very Short Introduction,* Oxford University Press, USA. Kindle Edition, 2013.
Close, Frank. *Particle Physics: A Very Short Introduction,* Oxford University Press. Kindle Edition, 2004.
Cowling, Graham; Allen, Terence. *The Cell: A Very Short Introduction*, Oxford University Press. Kindle Edition, 2011.

Feynman, Richard P. *Six Easy Pieces,* 1965.
Feynman, Richard P. *QED: The Strange Theory of Light and Matter,* Princeton University Press. Kindle Edition, 2014.

Hall, Alan. *Electron,* Kindle Edition, 2011.
Hall, Calvin. *Northern Lights, The Science, Myth and Wonder of Aurora Borealis*, 2001.

Hickey, Isabel M. *Astrology, A Cosmic Science,* SCB Distributors. Kindle Edition, 1992, 2011.

King, Andrew. *Stars: A Very Short Introduction,* Oxford University Press. Kindle Edition, 2012.

Lear, Benjamin & Shana. *Electron Orbitals and Electron Configurations* Kindle Edition, 2013.

Pederson, Daryl. *Northern Lights,*
The Science, Myth and Wonder of Aurora Borealis, 2001.
Polkinghorne, John. *Quantum Theory: A Very Short Introduction,* Oxford University Press. Kindle Edition, 2002.

Randall, Lisa. *Dark Matter and the Dinosaurs,* Harper Collins. Kindle Edition, 2015.
Redfern, Martin. *The Earth: A Very Short Introduction,* Oxford University Press. Kindle Edition, 2003.
Rothery, David A. *Planets: A Very Short Introduction,* Oxford University Press. Kindle Edition, 2010.

Sagar, Surendra Kumar. *Six Words,* Kindle Edition, 2014.
Scerri, Eric R. *The Periodic Table: A Very Short Introduction,* Oxford University Press. Kindle Edition, 2011.
Seymour, Percy. *Dark Matters: Unifying Matter, Dark Matter, Dark Energy and the Universal Grid*, 2008.
Stannard, Russell. *Relativity: A Very Short Introduction,* Oxford University Press. Kindle Edition, 2008.

Tuniz, Claudio. *Radioactivity: A Very Short Introduction,* Oxford University Press. Kindle Edition, 2012.

Yee, Jeff. *The Particles of the Universe,* Kindle Edition, 2012.

INDEXES

What follows are 3 indexes each with their own symbol for ease of identification:

- ☼ **Astrological Index** for astrological thinking within CPI Theory.

- ○ **Solar System Index** relating to activities within our solar system.

- ➢ **Electromagnetic Phenomena Index** for issues arising from electromagnetic interactions.

☼ Astrological Index

☼ A

- ☼ Actualized, 156
- ☼ Ahuja, Chand Karan, 51
- ☼ Angular Interaction, 325
- ☼ Apparent Motion, 65, 325
- ☼ Archetypal Wave Forms, 156
- ☼ Armstrong, Robin, 13, 18, 51, 59, 134, 157, 160, 167, 177, 190, 280, 281, 283
- ☼ Ascendant, 177
- ☼ Aspects, Major, 156
- ☼ Astro*Carto*Graphy, 10
- ☼ Astrologer, 9, 10, 17, 18, 30, 43, 51, 132, 134, 141, 156, 157, 160, 161, 164, 168, 184, 201, 228, 239, 240, 242, 251, 252, 280, 283
- ☼ Astrological Association of Great Britain, 325
- ☼ Astrological Birthchart, 239
- ☼ Astrological Chart, 140, 187
- ☼ Astrological Cusp, 42
- ☼ Astrological Elements, 20
- ☼ Astrological Energy Patterns, 135
- ☼ Astrological Implications, 59, 281
- ☼ Astrological Influence, 140, 167
- ☼ Astrological Insights, 160, 281
- ☼ Astrological Interpretations, 141
- ☼ Astrological Perspective, 140, 165
- ☼ Astrological Phenomena, 280
- ☼ Astrological Planet, 19, 22, 27, 34, 58, 160
- ☼ Astrological Planetary Energy Transference, 53
- ☼ Astrological Planetary Influence, 13
- ☼ Astrological Qualities, 165
- ☼ Astrological Signatures, 22, 42, 133, 135, 160
- ☼ Astrological Techniques, 9
- ☼ Astrological Theories, 325
- ☼ Astrologically Negative, 165
- ☼ Astrologically Positive, 165
- ☼ Astrology, 9, 11, 13, 15, 20, 22, 58, 65, 141, 157, 279, 281
- ☼ Astrology News Service, 10
- ☼ Attitudes and Patterns of Behavior, 157

☼ B

- ☼ Babylonian Astrologers, 10
- ☼ Biographical Correlations, 156
- ☼ Brady, Bernadette, 242

☼ C

- ☼ Cardinal, 166
- ☼ Cardinal Activity, 41
- ☼ Cardinal Atom, 165, 166
- ☼ Cardinal Electron, 165, 166
- ☼ Cardinal Ion, 166
- ☼ Cardinal Quality, 41, 165
- ☼ Collective Over the Individual, 280
- ☼ Cosmic Dust, 218
- ☼ Currey, Robert, 10, 18, 51, 134, 283

☼ D

- ☼ Destiny, 157, 160
- ☼ Destiny, Electron, 280

289

291

Solar System Index

293

297

298

➤ Electromagnetic Phenomena Index

A

F

312

316

317

320

S

About The Author

Tony Waterfall has been involved in many pioneering astrological efforts over the last 45 years. In 1975 he was Director of Registration for the World Astrological Directory. In 1995 Tony hosted 320 continuous weeks of the online "Astrology Hour" that started in the Matrix Forum on MSN.

Tony Waterfall spends his time working through and publishing astrological theories that include the linking of multiple charts; progressed/relocated, tropical/sidereal, geocentric/heliocentric using exactitudes; the exact degree timing of angular interaction and apparent motion; direct and retrograde motions.

Member: National Council for Geocosmic Research and Astrological Association of Great Britain.

Linkedin.com Tony Waterfall